QB
755
M435
1987

Mark, Kathleen.

Meteorite craters.

$9.95                                    17345

# Meteorite Craters

*Within Meteor Crater, Arizona*

# *Meteorite Craters*

## KATHLEEN MARK

**The University of Arizona Press**
TUCSON

*About the Author*

Kathleen Mark has long been interested in geology and in writing about it for general readers. Since the early 1970s she has published articles on a variety of geological topics. In 1975 she was a co-winner of the Nininger Award for a short work on meteorite craters. This book presents the first comprehensive, non-technical history of the recognition of meteorite craters on the earth.

THE UNIVERSITY OF ARIZONA PRESS

Copyright © 1987
The Arizona Board of Regents
All Rights Reserved

This book was set in ITC Bookman.
Manufactured in the U.S.A.

Library of Congress Cataloging-in-Publication Data

Mark, Kathleen.
  Meteorite craters.

  Bibliography: p.
  Includes index.
  1. Meteorite craters.   I. Title.
QB755.M435   1986        551.4'4        86-19244
ISBN 0-8165-0902-6 (alk. paper)

British Library Cataloguing in Publication data are available.

# Contents

# *Illustrations*

# *Preface*

THE DESIRE TO KNOW IS A distinguishing characteristic of the human species. It was through attempts to understand natural phenomena that primitive societies developed their mythical cosmologies, which in many cases were modified or discarded as sophistication increased. Some of the early Greeks attempted to reject mythology and to understand the world in terms of its physical properties. No doubt there were even earlier attempts. With excursions into the realms of both physics and metaphysics, the search for truth has continued. It has emerged in recent centuries in the form of modern science.

Pure science, concerned with the objective investigation of natural substances and processes, has often been criticized as materialistic. Nevertheless, objective investigation has shown that the materials which constitute the world and everything in it are marvelously complicated, and that they interact in a consistently predictable manner according to apparently universal "laws." This concept of natural phenomena, because it was derived from observation of materials, may indeed be described as materialistic. Its impact, however, has been intellectual, for it has profoundly affected our ideas concerning the earth and its inhabitants. In contrast to views that were prevalent several centuries ago, the earth is no longer regarded as having been created expressly for the benefit of humanity, nor as the stable centerpiece of a cozy universe consisting of the sun and the visible planets. Objective investigation has provided unassailable evidence indicating that the earth is a relatively small member of our solar system—a system that is immense by human standards, but is only one of many such systems in a vast universe.

In addition, investigation has shown that cosmic bodies, both within our solar system and beyond it, are composed of materials similar to those which constitute the earth, and are subject to the same laws of interaction. Evidence came from several lines of investigation. One of them was the study of meteorites—the bits and pieces (often small) of rocky or metallic material of cosmic origin that occasionally strike the earth. Chemical analysis of meteorites revealed that they contain some of the same elements, often in similar combinations, as those to be found in terrestrial rocks—indicating, apparently, that the materials of the earth are also the materials of the universe. Confirming evidence came from spectroscopic analysis of light from remote galaxies, which showed the presence of familiar elements at enormous distances—many light years—from the earth.

The study of meteorites commenced about two centuries ago. Study of the effect that the impact of large meteorites may have upon the earth is very recent, however, and still incomplete; but it is now known that the earth bears many scars which indicate collisions with large meteorites in the geologic past. It is at last agreed, after much controversy, that most of the craters on the moon were caused not by volcanism but by the impact of meteorites. Similar evidence of meteoritic impact has been observed on the other earth-like planets and on some of their moons. In fact, many authorities now suggest that the planets were formed by accretion of material supplied by a long succession of collisions with meteorites.

A remarkable change of opinion has occurred since the beginning of the twentieth century, when the idea that a meteorite crater might exist on the surface of the earth was given little serious attention. Opinion changed slowly at first, and with difficulty. A few holes were convincingly shown to be meteorite craters. A few more were discovered. Gradually the means for recognizing them were established, and since the mid-1960s many ancient impact sites have been located on the earth. It is the history of the objective investigations which eventually led to the recognition of terrestrial meteorite craters that this book attempts to trace.

Several years ago, for a short paper on this subject I was fortunate enough to be a co-winner of the Nininger Award. Without this encouragement the book would probably not have been written. I am deeply grateful to H. H. Nininger for his generosity, and to Carleton B. Moore, of the Center for Meteorite Studies, Arizona State University, for his part in administering this award.

I am indebted to a great many people whose comments and criticism have been tremendously helpful. Wolfgang E. Elston, of the Geology Department of the University of New Mexico, read an early draft of the manuscript and made valuable suggestions. In addition, he kindly supplied a number of anecdotes concerning Walter H. Bucher, his former professor at Columbia University, and allowed me to make use of some of Bucher's correspondence. Other members of the Geology Department of the University of New Mexico have always been available for questions and comments. In particular, I am grateful for the assistance of D. G. Brookins and of Jon Callender.

Robert S. Dietz read an early version of the manuscript and also read it in its final form. He has my sincere thanks for his encouragement, comments, and suggestions, which have been of very great help, and also for the generous manner in which he made his photographs available.

E. M. Shoemaker read portions of the manuscript, and I am most grateful for his suggestions, which have been incorporated.

Because of his investigations concerning the history of Meteor Crater, the late William Hoyt was an authority on the subject. After reading an early draft of the manuscript, he made comments and suggestions which were much to the point and greatly appreciated.

Others who read early drafts of the manuscript in various stages of its development are Carol Hill, Laurence Cohen, Joan and John Neary, Marke Talley, and Christopher Mark. My sincere thanks are offered to all of them.

I am deeply grateful, also, to Paul Fitzsimmons and to the late Bernd Matthias for translations of Russian and German papers.

I am very much indebted to many people who helped me to obtain illustrations, in particular to members of the Dominion Observatory, Ottawa, Canada, who kindly supplied photographs of some of the early-recognized Canadian meteorite craters. I am most grateful, also, to W. Reiff, who assisted me to obtain a photograph of the Steinheim Basin; and I would like to offer particular thanks to Francis J. Turner, who graciously rephotographed his microsections of Yule marble for inclusion in this book. Sincere thanks are offered to the many investigators who kindly granted me the use of their photographs; and my thanks, also, to Jason Grammer and Elaine Faust, who drafted several of the figures.

Finally, I thank my husband, Carson Mark, who read much of the manuscript piecemeal as it developed, and whose patience during the years while it has been taking shape has never worn thin. In particular I appreciate his pointing out the misprint in Gifford's table of energy in calories per gramme, which had escaped me entirely.

And if, in spite of the kindness and help of so many people, errors still remain in the book, they are my responsibility.

K. M.

# *Introduction*

IN THE EARLY YEARS OF THE twentieth century, an enormous bowl-shaped hole in the high plains of northern Arizona was investigated by a mining engineer and his associates. The hole was almost three quarters of a mile wide—a little over one kilometer[1]—and it penetrated thick layers of subsurface rock. The investigators concluded that it was created thousands of years ago when a mass of meteoritic iron struck the earth, and they produced extensive evidence in support of their claim.

This was a startling idea. To many geologists at that time it was preposterous. True, it was known that meteorites do sometimes strike the earth. In fact, many stony and metallic meteorites had been discovered in various parts of the world, but the ground where they fell had been little disturbed. Nowhere was a large hole known to have been created by the fall of a meteorite. The evidence presented by the investigators of the Arizona hole was not accepted by some of the most highly respected geologists in the country. In their opinion, the great hole was the result of volcanic processes.

The investigators remained convinced that their interpretation was correct, and a controversy developed that continued for more than two decades. As the Scottish geologist, James Hutton, remarked almost two centuries ago, it is often through controversy that truth is made to appear. This one attracted the attention of a number of other scientists, and several pits and depressions in widely separated parts of the world, suggested as possible meteorite craters, were critically examined. Some of them displayed marked similarities to the hole in the Arizona plains. Some had fragments of meteoritic material associated with them. And some were assumed to be meteorite craters because no other origin seemed likely.

As these localities were examined, much of the early skepticism disappeared. By the end of the 1920s the existence of meteorite craters on the surface of the earth was accepted by some people as an established fact, and the controversy concerning the Arizona crater, which by that time was known as Meteor Crater, had almost disappeared. (The word "crater," incidentally, was used in its general sense denoting a bowl-shaped depression with no implication as to its origin.) In the early 1930s, nine probable meteorite craters, randomly scattered over the globe, were authoritatively listed. Twenty years later the number had doubled. Today more than a hundred places are known where at some time in the past a meteorite struck the earth. The number continues to grow as new discoveries are made.

The question of recognition proved to be a difficult one. In fact, the search for unique criteria to distinguish meteorite craters from pits and depressions due to other causes became more complicated as more impact sites were discovered. It soon became clear that the early-recognized meteorite craters, most of which were roughly bowl-shaped, were of relatively recent origin. Estimates suggested an age for the Arizona crater of less than 100,000 years, and perhaps as little as 20,000—youthful indeed, geologically speaking. It was realized also that processes of change, always at work on the surface of the earth, must eventually obliterate such craters. They might be filled by drifting sands or water-borne sediments, or by slumping of their sides, or they might be completely worn away by regional erosion. Geologically ancient impact sites, therefore, would be unlikely to have the appearance of bowl-shaped holes. Many of them would no doubt have been completely erased; occasionally, however, revealing traces of an impact event might remain. What would such traces consist of? How could they be detected?

The scar created when a meteorite strikes the earth must depend to a large extent on the nature and velocity of the meteorite. It must also be affected by surface conditions on the earth at the point of impact—and whatever those conditions may be today, they were no doubt very different several hundred million years ago. During the 1950s a search for impact scars was undertaken in Canada, in rocks of immense age. It resulted in the discovery of many examples of what the Canadian investigators called "fossil" meteorite craters, and it solved some of the problems of detecting them. It also showed, as had been suspected, that the form of ancient meteorite craters as observed at the surface of the ground is much less uniform than the bowl-shaped holes created by relatively recent impact events. Evidence was discovered, in fact, suggesting that ancient meteorites had occasionally pierced the crust of the earth. By the early 1960s, threads of information from the investigators of both ancient and recent meteorite craters, and also from researchers in other scientific disciplines, were woven into a fabric of solid fact concerning the structure of meteorite craters and the means for recognizing them.

Increased understanding of meteoritic impact had some interesting and unexpected results. One of them was to end a long and rancorous controversy concerning the markings on the moon. For many decades it was believed that lunar craters resulted from volcanism, an opinion that was challenged from time to time by claims that they were created when meteorites struck the moon. During the 1960s it was shown to the satisfaction of most scientists that a large proportion of the formations on the surface of the moon had indeed been caused by meteorite impact. Later this was confirmed by examination of the moon-rock samples brought back from the lunar surface by the astronauts. And eventually, space explorations revealed that meteorite craters exist not only on the earth and the moon, but also on the other inner planets of the solar system—Mars, Mercury, and probably Venus—no doubt indicating similar histories and development.

It is now known that the earth bears many scars which record violent encounters with large meteorites in the geologic past. The work at the Arizona crater at the beginning of this century was the first of a long series of investigations, which often seemed unrelated at the time, and which eventually led to greatly increased understanding of meteoritic craters and other meteoritic phenomena. It is interesting to note that this work began almost exactly a century after European scientists finally accepted the fact that now and then meteorites actually do strike the earth. Before that time, reports of stones or metal falling from the sky were discounted as superstition.

## CHAPTER 1

# The Recognition of Meteorites

MOMENTARY LUMINOUS STREAKS in the sky—shooting stars, or meteors—occur when metallic or stony objects originating in interplanetary space enter the earth's atmosphere and their surfaces are heated to incandescence by friction with the air. Such objects range in size from dust particles to masses weighing many tons. Many of them are destroyed by friction-generated heat during their passage through the air. Those which survive and reach the earth are known as meteorites.

Because meteorite falls are rare and unpredictable, any observation of them takes place quite by chance. However, accounts of spectacular events accompanying the falling of stones from the sky are to be found in the early histories of many regions. Such things must occasionally have been witnessed even in prehistoric times, for meteorites have been found among relics of ancient societies in circumstances suggesting that they were revered, possibly even worshipped. In other cases, meteoritic iron evidently proved to be extremely useful, for it was fashioned into knives and other instruments. It has been suggested that the first iron used by human beings was meteoritic.

In the eighteenth century the idea that stones or pieces of iron could fall from the sky was considered absurd by most European scientists. It must be remembered that investigators were attempting to establish scientific truths through observation, and to distinguish fact from superstition; but even in the midst of such worthy intentions, preconceived notions were difficult to dislodge. The numerous accounts in ancient writings of the falling of stones from heaven or from the clouds were regarded with considerable skepticism, which was not lessened by the fact that they were sometimes associated with miraculous rains of other improbable things such as milk, blood, iron, flesh, wool—even baked bricks.[1] Similar miracles were recounted at least as late as the sixteenth century, when freak actions of lightning were said to be accompanied by fourteen prodigious kinds of rain, including stones, iron, flesh, fish, worms, and frogs. There were even reports of showers of fire and brimstone. Another remarkable downpour was described in 1683 by Robert Plott, a professor of chemistry at Oxford, who wrote of the occurrence of a rain of frogs.[2]

Few people, however, actually witnessed these miraculous rains. Similarly, few people witnessed the falling of meteorites, and in the late eighteenth century those who claimed to have done so were rebuked for their credulity. Accounts of meteorite falls frequently bordered on the fantastic and did little to add to the credibility of the phenomenon. Folklore on the subject included the appearance of fiery serpents in the sky, sometimes accompanied by other apparitions, along with plentiful fire, smoke, thunder, and even earthquakes. To many European scientists in the second half of the eighteenth century, the falling of stones from the sky was a physical impossibility, and they did not hesitate to say so.

Brilliant meteors—fire-balls, or bolides—were nevertheless observed from time to time. Occasionally, also, it was claimed that stones had been seen to fall from heaven without the accompaniment of celestial fireworks. Descriptions of both kinds of event found their way into the scientific literature, along with speculations as to their cause. For instance, a spectacular light was seen in the sky by many people in England on September 20, 1676, and was described by John Wallis as follows:

> In the dusk of the Evening (about Candle-lighting) there appeared a sudden light, equal to that of Noon-day; so that the smallest pin or straw might be seen lying on the ground. And, above in the Air, was seen (at no great distance as was supposed) a long appearance as of fire; like a long arm (for so it was described to me) with a great knob at the end of it; shooting along very swiftly: and, at its disappearing, seemed to break into small sparks or parcels of fire, like as Rockets and such Artificial Fire-works in the Air are wont to do. 'Twas so surprising, and of so short continuance, that it was scarce seen by any who did not then happen to be abroad. 'Twas judged, by him from whom I first heard of it, (for I had not the hap to see it myself,) to continue about two or three minutes: But, I find he took a minute to be a very short time, (little more than a moment.) From others I am told, it was scarce longer than while one might tell fifteen or twenty at the most; which will be less than half a minute. . . . But that which makes it to

me the more surprising, is this; that I find the same to have
been seen in most parts of *England,* and at or near the same
time. . . .[3]

After listing a dozen or so counties in England in which the meteor
was seen, he continues:

In how many other parts of *England,* or in what parts out of
*England* it might be seen; I have not yet heard. But this is a
great breadth of ground, and too much for an ordinary Meteor
in our lower region of the Air to be seen in at once: Yet (for ought
I hear) it is agreed by all to have been seen at the same time,
between seven and eight at night the same day, in the dusk of
the Evening. Which argues, that either it was higher than
they imagined, (though the light of it reached the Earth) or
else, that it had a very swift motion. This made me then
conjecture . . . that it might be some small Comet. . . .[4]

The astronomer Edmund Halley was similarly perplexed by the
phenomena of meteors, or "lights in the sky." In his time such things
were attributed somewhat vaguely to accumulations of gases in the
atmosphere, although Halley pointed out "that at 40 Miles high the
Air is rarer than at the Surface of the Earth about 3000 times; and
that the utmost Height of the Atmosphere, which reflects Light in the
*Crepusculum* [twilight], is not fully 45 Miles."[5]

Halley agreed with Wallis that the great light seen over England in
1676 must have been "very many Miles high." He calculated, also,
that the meteor seen at many places in England on July 31, 1708, was
"evidently between 40 and 50 Miles perpendicularly high" and that it
was moving with incredible velocity. Halley mentioned other examples. One was a meteor which appeared over Germany in the year
1686 and was described by Gottfried Kirch (known as a diligent
observer of the heavens), who noted that this meteor was seen by
persons in widely separated localities and must therefore have been
exceedingly high above the surface of the earth. Another was the
famous meteor which passed over Italy on March 21 (old style) 1676,
observations of which suggested a similarly great altitude; its astonishing velocity, it was said, "could not be less swift than 160 Miles
in a Minute of Time." In addition, its passage was accompanied by a
hissing noise and was followed, after it went out beyond the coast in
the direction of Corsica, by a tremendous blow as it was extinguished
in the sea. Then there was another sound like the rattling of a great
cart running over stones, a noise which continued, according to the
description by Professor Geminian Montanari, for "about the time of
a *credo.*"

In his report to the Royal Society, Halley remarked:

It may deserve the Honourable Society's Thoughts, how so great a Quantity of Vapour should be raised to the very Top of the Atmosphere, and there collected, so as upon its Accension [ignition] or otherwise Illumination, to give a Light to a Circle of above 100 Miles Diameter, not much inferior to the Light of the Moon; so as one might see to take a Pin from the Ground in the otherwise dark Night. 'Tis hard to conceive what sort of Exhalations should rise from the Earth, either by the Action of the Sun or subterranean Heat, so as to surmount the extream Cold and Rareness of the Air in those upper Regions: But the Fact is indisputable, and therefore requires a Solution."[6]

Halley suggested an ingenious solution a few years later, in his account of the meteor of 1718, which was "seen over all parts of the Islands of Great Britain and Ireland, over all Holland and the hither parts of Germany, France, and Spain at one and the same instant of Time." It was thought that certain "inflammable sulphureous Vapours" had an "innate Levity," and therefore might ascend from the earth's surface to heights far above the reputed limits of the atmosphere. They might, Halley suggested, eventually "lie like a Train of *Gunpowder* in the *Ether*," and through some internal ferment similar, perhaps, to the damps in mines, they might eventually catch fire; and if this were the case, a meteor would not be a moving globe of fire, but "a successive kindling of new Matter."[7]

Halley's "solution" was challenged some years later. A painstaking study of the meteor seen in England and Scotland on November 26, 1758, was made by John Pringle, a distinguished physician with scientific interests. He was widely acquainted, and collected reports of the meteor from many different localities. From these reports he carefully determined its path and estimated its velocity to have been "above 100 times swifter than the mean celerity of a cannon ball, and nearly equal to that of the earth in its orbit around the sun." He found the numerous accounts he received to be in rather general agreement as to the intensity of the light, the emission of sparks, and the brief duration of the event. Some people reported a final break-up or explosion of the fire-ball; others did not. The final thunderous noise, even though in some instances it was not perceived, was mentioned by many people and led him to the conclusion that "the substance of the meteor was of a firmer texture than what could arise from mere exhalations, whether formed into a sphere and then burning; or disposed into a kind of train and consumed by a running fire." Among other arguments for the solidity of the body, Pringle listed its extreme velocity and the intensity of its light as "circumstances more conformable to a heavy and solid substance than to one formed of exhalations only."[8]

A somewhat similar criticism of Halley's suggestion appeared in 1784 in "an account of some late fiery meteors" by Charles Blagden:

> Dr. Halley gives no just explanation of the nature of these vapours, nor of the manner in which they can be raised up through air so extremely rare; nor, supposing them so raised, does he account for their regular arrangement in a straight and equable line of such prodigious extent, or for their continuing to burn in such highly rarefied air. Indeed, it is very difficult to conceive, how vapours could be prevented, in those regions where there is in a manner no pressure, from spreading out on all sides in consequence of their natural elasticity, and instantly losing that degree of density which seems necessary for inflammation. Besides, it is to be expected, that such trains would sometimes take fire in the middle, and so present the phaenomenon of two meteors at the same time, receding from one another in a direct line.[9]

Blagden listed various observations of the meteor which occurred on August 18, 1783, and was seen in many parts of Scotland and England and also at Dunkirk, Brussels, and Paris. They indicated, in quite remarkable agreement, that the height of the meteor must have been more than fifty miles, and, as for its velocity—"at the lowest computation of 20 miles a second, it exceeds that of sound above 90 times, and begins to approach that of the earth in her annual orbit."[10] Blagden referred to accounts of the aurora borealis in Siberia and other places in the far north and to the astonishing velocity with which its spectacular streams of light traveled through the sky. They were said to be accompanied by hissing, crackling, and rushing sounds, similar to those often attributed to the passage of meteors. Meteors, in his opinion, were probably a manifestation of electricity related in some way to the northern lights.

In 1776 a Jesuit priest, D. Troili, who had made a study of chemistry, reported having witnessed the fall of a stone near Albaretto, Italy. His account received little attention at the time. However, an iron sulfide which he discovered in this stone, and which was later found to be an important component of many meteorites, is now known as troilite.

In 1768 a stone was seen to fall from the sky by another priest, the Abbé Bachelay. It was said to have been picked up while still warm. In due time it was presented by the Abbé to the Royal French Academy, where the account of its fall was discredited. It was pronounced to be just an ordinary stone—merely a kind of pyrite (an iron sulfide). Its peculiar black fusion crust was said to be the result of its having been struck by lightning. Members of the Academy similarly discredited accounts of the origin of the stone of Ensisheim, in Alsace, reported to have fallen from the sky in 1492. Since that time it had been kept in a church, securely fastened by a chain. It, too, was just an ordinary stone, they said, and the story of its fall should be attributed to ignorance and superstition.

In spite of complete lack of interest on the part of most recognized scientists of the day, many other accounts of both meteors and the falling of stones from the sky existed in the scientific literature. These accounts, along with all available samples of fallen material, were exhaustively studied by E. F. F. Chladni, a physicist connected with the University of Berlin. Chladni published a book in 1794 in which he developed a theory linking the two phenomena—that of meteors sometimes accompanied by the falling of stony or metallic masses, and that of the falling of such masses without the simultaneous appearance of fiery streaks in the sky—as resulting from the same cause. This book marked the beginning of the science of meteoritics.

A fire-ball, or bolide, is a natural phenomenon, said Chladni, in effect, thus ruling out the fantastic and the supernatural. At its first appearance it looks like a bright star, or a shooting star, increasing in size as it comes down until it may appear as large as or larger than the full moon. Sometimes it emits smoke, flames, and sparks, and it finally explodes with a loud bang. As several other observers had done, Chladni regretted the fact that persons with astronomical training had seldom observed a fire-ball. He urged chance witnesses of such events to make careful observations of trajectory, shape, size, color, altitude, velocity, accompanying sounds, and time of day.

Chladni declared that the parabolic trajectories of fire-balls indicated that they were affected by gravity. He agreed with Pringle and other observers who claimed that they must be of firmer texture than gas. In fact, Chladni maintained that they must be composed of dense, heavy material. As such, however, they could never have been raised by any known earthly force to the heights at which they are observed. Fire-balls must therefore come to us not from the earth but from the universe.

A number of suggestions as to the origin of fire-balls were current at that time among the few scientists who had any interest in them. Chladni demolished these in short order. The idea that fire-balls were some sort of emanation from the aurora borealis because they traveled from north to south was not justified. Many were seen coming from other directions. Nor could the explanation be that they were due to the transition of electricity from a place where it was in excess to a place where it was deficient, because in the near-vacuum of very high altitudes electricity would be diffused. It would appear in vague shapes similar to those of the aurora borealis rather than in the well-defined shapes and sizes presented by fire-balls. Some authorities considered lightning to be the origin of fire-balls. This could not be the case, said

Chladni, because there were many instances of fire-balls occurring in a cloudless sky with no accompanying lightning. He rejected the idea of a collection of inflammable vapors, because loose vapors would not be capable of the rapid motion exhibited by fire-balls and would not stand rapid and continuous burning.

The suggestion Chladni considered most likely was another idea from Halley, who said in connection with the meteor which passed over Italy in 1676:

> . . . it cannot be wondered that so great a Body moving with such an incredible Velocity through the Air, though so much rarified as it is in its upper Regions, should occasion so great a hissing Noise, as to be heard at such a Distance as it seems this was. But 'twill be much harder to conceive, how such an *impetus* could be impressed on the Body thereof, which by many degrees exceeds that of any Cannon Ball, and how this *impetus* shou'd be determined in a Direction so nearly parallel to the Horizon; and what sort of Substance it must be, that could be so impelled and ignited at the same time: there being no *Vulcano* or other *Spiraculum* of subterraneous fire in the N.E. parts of the World, that we ever yet heard of, from whence it might be projected.
>
> I have much considered this Appearance, and think it one of the hardest things to account for, that I have yet met with in the *Phaenomena* of *Meteors*, and am induced to think that it must be some Collection of Matter form'd in the *Aether*, as it were by some fortuitous Concourse of Atoms, and that the Earth met with it as it past along in its Orb, then but newly formed, and before it had conceived any great *Impetus* of Descent toward the Sun. . . .[11]

Halley, in other words, concluded that meteors were cosmic bodies. Chladni noted that this opinion was shared by the astronomers Maskelyne and Hevelius and the chemist Troili, as well as himself.

The theory of these natural phenomena, said Chladni, in effect, must be this: Our earth is composed of earthy, metallic, and other materials, including a large proportion of iron. It seems likely that other celestial bodies are also composed of these materials. Small masses of them resulting perhaps from the breaking up of stars may be scattered about the universe. They have no connection with any major celestial body, but when they come within the gravitational range of a larger body, they will move in upon it. Such relatively small masses reach the earth as meteors or fallen stones.

Chladni searched the scientific literature, listing a great number of accounts of the fall of stony and metallic masses, and he carefully examined and described all samples of allegedly fallen material that he was able to obtain. For instance, a piece of stone from Eichstädt, which fell with a loud thunderclap near a hut where bricks were made, was found to consist of a thin skin of pure iron enclosing an ash-gray sandstone-like rock, throughout which were sprinkled small grains of pure iron and yellow-brown iron oxide. The rocks of the mountainous region where the fall occurred consisted of coarse marble, hornblende, chalk, and sandstone. It seemed probable that the fallen stones had no connection either with the surrounding mountains or with lightning. But it was unfortunate, as Chladni pointed out, that no investigation was made as to whether there was a real thunderstorm at the time, or whether iron was to be found in the immediate vicinity.

Taken all together, the various accounts added up to a convincing argument, if not proof, that stones do actually fall from the sky. Chladni's compilation also provided a general description of the character of such stones. A few of his examples will suffice here: The Arab philosopher Avicenna (980–1037) saw a stone containing sulfur fall from the sky in Cordova, Spain. In 998, accompanied by the sound of thunder, a stone fell into a field close to the River Elbe in Germany, and another stone fell in a nearby city. In 1559, near a range of mountains in eastern Germany, five stones about the size of skulls (presumably human skulls) fell from the sky accompanied by a smell of sulfur and a terrific thunderclap. In March 1631, in Silesia, a huge stone came out of a clear sky with one big bang; it was covered by a skin that appeared to be charred, and inside it looked like ore and was easily crushed. In 1698 a black stone fell in Berne with much noise. About the year 1700, in Jamaica, a ball of fire fell with a great blaze; the ground was broken, the grass was burned, and there was a smell of sulfur. In 1723, in Bohemia, out of a sky which contained one little cloud but was otherwise clear, more than thirty stones fell with a great deal of noise but no lightning. The stones were black on the outside, and inside they looked like ore. Chladni listed many more examples of fallen stones. In fact, during the late 1700s, in spite of official skepticism, increasing interest in such matters was reflected in the increasing number of observed and reported falls.

During an exploring expedition in Russia in the years 1768–1774, the German naturalist Peter Simon Pallas collected a wealth of previously unrecorded information about the vast Russian Empire. Among other things he determined the rock succession in various mountains, discovered fossil elephants in Siberia, and was shown a large mass of iron near the Yenesei River between Krasnojarsk and Abaskansk. The weight of the iron mass was estimated to be about 1,500 pounds. Pallas did not believe the claims of local inhabitants that it had fallen from the sky, but he was interested in its curious structure and soon became convinced that it was of natural origin and not a product of smelting. He arranged to have the iron transported to St. Petersburg (now Leningrad), where it became known as the Pallas Iron and, incidentally, was the beginning of a famous collection of meteorites.

Chladni studied samples of the Pallas Iron and described it as pure malleable metal with a sponge-like form. It was covered with a stone-like skin of iron in which were many somewhat spherical depressions. The cavities within it contained a hard amber-colored glassy material, identified by Pallas as olivine (a mineral now known to be a common component of many meteorites). The iron mass had been discovered on the ground, high in slate mountains. Some iron mines existed nearby, but they were at a lower altitude. The Pallas Iron was quite unlike clinkers produced by metallurgical processes; such clinkers are likely to be black and opaque and never contain pure iron. It could not be a product of smelting because it would then have been hard and brittle instead of malleable. According to Chladni, Pallas pointed out that the old Siberian miners, traces of whose work could still be found, seemed not to have had an iron culture, since their cutting tools were made of copper and bronze. The characteristic clinkers resulting from their smelting processes were quite different from those which would have been created by the smelting of iron ores. And in any case, if the iron had been the product of some metallurgical process, why would it have been abandoned high in the mountains?

Occasionally the fall of metal had been witnessed. In Bohemia, on July 3, 1753, material described as shiny, refractory iron ore embedded in a greenish mineral fell from the sky with a noise like thunder. Its surface was a layer of pure iron. Chladni felt that further investigation of it was highly desirable.

An isolated metallic mass was reported in 1773, when several tons of pure, malleable iron, of sponge-like form, were dug from beneath the cobblestones of the city of Magdeburg. A piece of it was cut off and forged, and was then hardened and polished. The result could be compared to the best English steel.

Another isolated mass of pure malleable iron, estimated to weigh several metric tons, was discovered in 1783 by Rubin de Celis. It was found in South America, partly buried in a chalk plain. It, too, exhibited indentations on its outer surface, which was covered by a skin of iron ochre, a yellowish iron oxide. Beneath it, farther down in the earth, there was no trace of iron. Within a hundred miles there were no iron mines and no mountains.

Chladni gave detailed accounts of various other isolated masses of pure, malleable iron which were so similar to one another that they must have formed under similar conditions. Because of their structure, their purity, and their malleability they could not have been products of smelting. They could not, as some people suggested, be of volcanic origin, for if that were the case the iron would inevitably contain impurities. Like the stony masses which were also strikingly similar to one another, they must have fallen from the sky.

Because of his work in acoustics, Chladni was a respected scientist, and his book, published in 1794, received some attention. At about the same time, interest was further stimulated by the occurrence of three well-documented meteorite falls. Stones from these and other falls were carefully examined. The results, along with Chladni's well-reasoned arguments, brought about a change in official opinion.

One of the falls occurred in Siena, Italy, in 1794. During a violent thunderstorm which took place eighteen hours after an eruption of Vesuvius, a dozen or so black stones of various weights and dimensions fell from the sky. They were quite different from any stones to be found in nearby territories. Their surfaces were black, and they showed signs of having been exposed to great heat. Inside, they were of a light gray color and contained shining spots which were said to be pyrite.

Another fall occurred on a mild, hazy day in December 1796. Near Wold Cottage in Yorkshire, England, explosions were suddenly heard, and a stone fell into a field where it penetrated the surface, throwing up quantities of earth. The stone was warm, smoking, and smelled of sulfur. It weighed some 56 pounds, and it was unlike any stone to be found in the region. The nearest volcano, it was pointed out by observers, was Hekla, in Iceland.

The third fall was near Benares, India, where on December 19, 1798, a luminous meteor appeared in a clear sky. The fall of a number of stones, accompanied by a noise like thunder, was witnessed by many people, including several Europeans. The stones were covered with a hard black incrustation and bore no resemblance to the common stones of the country. Inside they consisted of a whitish, gritty substance, which crumbled easily. Embedded in this material were small, spherical slate-colored globules (later known as chondrules), interspersed with particles of metallic iron. And it was noted that no active volcanoes were known to exist in all of India. It is interesting, incidentally, that volcanoes were mentioned in each of these reports, indicating that the idea that falling stones originated as debris tossed from erupting volcanoes, or were generated in some combination of ashes and clouds during volcanic eruptions, was current at this time.

The president of the Royal Society of London, Sir Joseph Banks, was a man of wide-ranging curiosity who, in the words of a contemporary, was "ever alive to the interest and promotion of science."[12] He was extremely interested in the accounts of various meteoritic events and collected all possible records of them. When the stone from Wold Cottage in Yorkshire was exhibited in London soon after its fall, the usual objections to the falling of stones from the sky were presented by members of the Royal French Academy. Banks, however, obtained a piece of the Yorkshire stone and observed that it was strikingly similar to a sample from the Siena fall which had been sent to him by a friend. He requested a young chemist, E. C. Howard, recently elected to the Royal Society, to examine the Yorkshire and Siena samples. Howard also received a sample of the Benares fall from a member of the Royal Society who lived in India, and a well-known mineralogist, Charles Grenville, loaned him a piece of the material said to have fallen in Bohemia in 1753—the very iron and stony material for which Chladni had felt further investigation to be desirable.

"Being thus possessed of four substances, to all of which the same origin had been attributed," Howard remarked in 1802, "the necessity of describing them mineralogically did not fail to present itself."[13] He pointed out that Chladni had "collected almost every modern instance of phenomena of this nature." He did not, however, accept Chladni's theory of the extraterrestrial origin of meteorites. "Whence their origin, or whence they came," he wrote, "is yet, in my judgment, involved in complete obscurity."[14]

Howard was highly critical of the methods of investigation employed by the Royal French Academy, whose analyses, as he pointed out, were "unfortunately made of an aggregate portion of the stone and not of each substance irregularly disseminated through it. The proportions obtained were, consequently, as accidental as the arrangement of every substance in the mass."[15]

The stone seen to fall in 1768 and presented to the Academy by the Abbé Bachelay was described by the academicians as covered on the outside with a thin black material which had apparently been affected by melting. Inside, its substance was of a pale cindery gray which, when examined under a lens, was found to be sprinkled throughout with an infinite number of brilliant metallic specks of a pale yellow color. A portion of the stone was ground up by the academicians and was found to be composed of sulfur, iron, and "vitrifiable earth" (i.e., silica). A sample of the Ensisheim stone was similarly ground up. It contained sulfur, magnesia, alumina, lime, and silica. These methods revealed nothing unusual. It was concluded, as we have seen, that they were ordinary stones which had been struck by lightning, possibly indicating that lightning was somehow drawn to pyritical material.

In comparing the four samples, Howard found that they differed in certain characteristics such as grain size, the size of the pyrite particles, the presence or absence of small spherical globules, and the proportion of metallic iron. Count de Bournon, one of the foremost mineralogists of the time, furnished Howard with a mineralogical description of them. He concluded that these stones, although they have not the smallest analogy with any of the mineral substances already known, either of a volcanic or any other nature, have a very peculiar and striking analogy with each other. This circumstance renders them truly worthy to engage the attention of philosophers; and naturally excites a desire of knowing to what cause they owe their existence. [16]

Howard made chemical analyses of the constituent parts of the stones, carefully separating the black coatings, the pyrite grains (when they could be seen), the spherical bodies (when present), the grains of metallic iron, and the earthy matrix. He found that the four stones, although they came from widely separated regions, had precisely the same character. They were composed of varying amounts of silica (50 to 75 percent), smaller amounts of magnesia and oxide of iron, and very small amounts of oxide of nickel. "The weight of nickel," he remarked, "is a mere estimation. We are not yet sufficiently acquainted with that metal to speak of it with accuracy, except as to its presence."[17]

The presence of nickel was of great interest, for it is seldom found in terrestrial rocks except in trace amounts. Nickel had already been discovered in the South America iron by J. L. Proust in 1800, but it had not previously been found in meteoritic stones. A sample of the South America iron was sent to Howard from the British Museum, and he confirmed Proust's results with, as he said, "great satisfaction in agreeing with a chemist so justly celebrated as Mr. Proust."[18] His chemical examination of other samples, including the Pallas Iron and iron contained in the Bohemia meteorite, revealed that they, too, contained nickel.

In 1802 Howard visited L. N. Vauquelin in Paris. Vauquelin had completed chemical investigations of several meteorites, and his results, though not yet published, confirmed those of Howard. Soon after, M. H. Klaproth in Berlin announced that he had undertaken similar work, with similar results. Vauquelin and Klaproth were at that time among the most highly respected chemists in Europe, and their endorsement of Howard's work carried great weight.

A. F. Fourcroy, in 1803, remarked that the attention of philosophers was much excited by the novelty of these results, and that some authorities considered the fallen stones to be "volcanic products projected from the moon."[19] And he noted a recent and dramatic example of the falling of stones from the sky:

At the time when we were most occupied with this new problem of natural philosophy, and while, uncertain in regard to its existence, we were discussing the authenticity of the accounts given of it by the antients [sic] and moderns, the inhabitants of Laigle and of a vast extent of surrounding district were witnesses of the phaenomenon: it appeared over their heads on the 26th of April [1803] with circumstances capable of striking with terror and astonishment. [20]

The spectacular fall which occurred near L'Aigle[21] in northern France was witnessed by a great many people. A member of the Paris Academy, J. B. Biot, was directed to investigate, and he confirmed the reports of the event as authentic:

The meteor did not burst at Laigle, but at the distance of half a league from it. I saw the awful traces of this phaenomenon; I traversed all the places where it had been heard; I collected and compared the accounts of the inhabitants: at last I found some of the stones themselves on the spot, and they exhibited to me physical characters which admit of no doubt of the reality of their fall. . . . The inhabitants say that they saw them descend along the roofs of the houses like hail, break the branches of the trees, and rebound after they fell on the pavement. . . . In this account I have confined myself to a simple relation of facts . . . and I shall consider myself happy if . . . I have succeeded in placing beyond a doubt the most astonishing phaenomenon ever observed by man. [22]

It has often been said that the well-documented L'Aigle fall brought about official recognition of the fact that meteorites exist, and that they do sometimes fall upon the earth. But due to the chemical analyses undertaken by Howard, Vauquelin, and Klaproth, that change of opinion had already occurred. As D. W. Sears has pointed out, rather than bringing it about, the L'Aigle fall served to underline the change of opinion that resulted from the chemical investigations. [23]

No sooner was the existence of meteorites officially recognized than interesting observations were made about them. G. Thomson, an Englishman living in Naples, was engaged in some experiments concerning the malleability of meteoritic iron.[24] In attempting to remove some obstinate rust spots from a sample of Pallas Iron which had been in his possession for several years, he used dilute nitric acid:

the nitric acid revealed to me the laminar and crystalline texture of the Siberian iron. At a suitable moment I arrested the solution of the metal in the diluted nitric acid and made it possible to distinguish, both by their different degrees of solubility and by their differing grades of polish, the three varieties of which it was composed. [25]

The three varieties were an alloy of iron and nickel, known as taenite; an alloy with a relatively low nickel content, called kamacite; and plessite, a mixture of the two.

Crystal patterns in the Pallas Iron were very small and indistinct, and investigation of them was hard on the eyes. Nevertheless, Thomson's report included a drawing of them sufficiently enlarged for the complicated structure to be clearly seen. He noted that the widely held opinion that a crystalline metal must be brittle was not correct, for a malleable metal, such as that of meteorites, could be crystalline. His work apparently escaped general attention, although an account of it was published in the *Atti dell' Accademia delle Scienze di Siena* in 1808. It was later shown, however, that a letter which was the basis of this publication was written in 1804.[26]

Also in 1808, crystal patterns in iron meteorites were observed by Alois von Widmanstätten of Vienna. He, too, chanced upon the use of acid (though there are indications that he may have discovered the patterns by annealing the cut surface of a meteorite[27]). His interest was aroused when a friend, Carl von Schreibers, director of a natural history museum, gave him a sample of the meteorite which fell at Hraschina, Yugoslavia, in 1751. His excellent illustrations (nature prints) of these patterns, which have become known as Widmanstätten patterns, were distributed among numerous friends. For years he studied the peculiar crystal structure found in some meteorites, so that its connection with his name is felt to be fully justified.[28]

*1.1   Portion of the Krasnojarsk (Pallas) iron meteorite, showing Widmanstätten structure; reproduced from G. Thomson, 1808. From F. A. Paneth (1960), who remarked that "it seems to be enlarged about five times."*

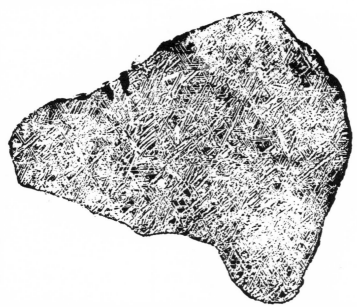

**1.2**  *Reproduction of a "nature-print" of the Elbogen meteorite. Widmanstätten's "nature-prints" were produced by rubbing the etched surface of the meteorite with printer's ink and using it as a printing block.*

During the nineteenth century the study of meteorites continued chiefly among mineralogists and chemists. Three general classes of meteorites were recognized—irons, stones, and stony irons—and investigations were undertaken to determine their chemical constituents. There was agreement with Chladni's contention that meteors and the falling of iron and stony masses, with or without accompanying fiery displays, are aspects of the same class of natural phenomena. This conclusion was apparently reached independently, for Chladni's work seems not to have been widely known. J. N. Lockyer, for instance, remarked in 1891:

> . . . in the thousands of records of the phenomena presented by luminous meteors, fire-balls, bolides, and shooting or falling stars as they have been variously called, we have the links which connect in the most complete manner the falls of actual irons and stones from heaven with the tiniest trail of a shooting or falling star. . . . The heaviest masses fall by virtue of their substance and compactness resisting the friction of the air, the smaller bodies are at once burnt up and fill the upper regions of the earth's atmosphere with meteoritic dust. The difference is only one of size. . . .[29]

Most of the meteorites found from time to time were discovered on the surface of the ground or buried beneath a few feet of it. When seen to fall, they were found in unimpressive holes at most a few feet deep, which had been created by their impact. The idea that a large crater could be formed by the falling of a meteorite was a startling one. It was not until almost the end of the nineteenth century that the study of such a crater, thought by a few people to be of meteoritic origin, was undertaken.

## CHAPTER 2

# Curious Landforms of the Colorado Plateau

EXCEPT ON THE PART OF A FEW specialists, interest in meteorites was sporadic during the nineteenth century, though it often increased briefly after some meteoritic event. In the United States, excitement flared up after notable meteorite showers—particularly one in 1833, seen in many parts of the country, when meteorites were said to have fallen like snowflakes. There was little sustained interest, however, on the part of the general public. As for meteorite craters, not only were they unknown, but the possibility of their existence was scarcely even considered.

The great hole in the Arizona plains, now known as Meteor Crater, lies within the Colorado Plateau, a geologically unique region of some 150,000 square miles that includes parts of Utah, Colorado, Arizona, and New Mexico. The region contains many peculiar and puzzling rock structures, many of which are of volcanic origin, and so it is hardly surprising to find that early interpretations of Meteor Crater attributed it to volcanism.

The vividly colored and curiously sculptured landforms of this semi-arid part of the country were studied by early geologists who traveled through the southwestern United States as members of government-sponsored exploring expeditions. They described many remarkable features of the Plateau country, and their observations contributed to the development of basic concepts concerning the evolution of landforms. It was clear to them that due to some unexplained cause the whole region had been subjected to gentle uplift for millions of years. Rapid run-off following infrequent but violent storms had carried away vast quantities of sand and rock fragments, eroding the high plateaus and carving out the canyons. All the early

**2.1  Index map of the Colorado Plateau.**

geologists noted that immense amounts of material had disappeared. This was evident, for example, in the fact that the steep sides of high plateaus rising on either side of a wide plain often displayed precisely the same succession of rock layers, and the same layers could be seen in sculptured spires and pinnacles which in some places rose abruptly from the plain. The plateaus and spires were evidently once connected, forming a single unit. The connecting material had been removed by erosion.

C. E. Dutton, of the United States Geological Survey, recognized spires and pinnacles of another type as volcanic "necks," composed of hard volcanic rock and debris. They had evidently formed with the cessation of volcanic action, when cooling lava congealed within vents through which volcanic extrusions once reached a ground surface hundreds of feet higher than that of the present. The upper layers of regional rock through which the hot volcanic material forced its way have been removed by erosion along with the extruded lava. All that now remains of the ancient volcanoes are hardened spires of volcanic rock—the volcanic necks. And these are not the only evidence of volcanism. Dutton, traveling through the southern part of the Colorado Plateau in 1885, described the constant appearance of lava:

> Sometimes it mantles the soil of a valley bottom, sometimes it is the capsheet of some mesa. It is scattered about in an irregular way, as if the molten stuff had been dashed over the country from some titanic bucket, and lies like a great inky slop over the brightly colored soils and clays. . . .[1]

Rock formed by volcanic extrusion is estimated to cover some 15,000 square miles of the Colorado Plateau. In Arizona, volcanic fields both north and south of the Grand Canyon contain numerous lava flows and many volcanic necks. And there are many walls of once-molten rock that cut across the rocks adjacent to them. These are known as dikes, though few of them are as high as that described by Dutton:

> It is no uncommon thing in the heart of the Plateau region to come suddenly upon a long narrow wall of black rock projecting hundreds of feet in the air, rising out of a flat plain. The rock is a dike of basalt or andesite; but the dike itself is all the eruptive material we see—no lava stream overflowing the adjoining plain, no bed of volcanic rubbish. . . .[2]

Several smoothly rounded mountains in the region were interpreted by another outstanding geologist, G. K. Gilbert, as created by the welling up of a body of lava. Instead of erupting at the surface of the ground, the lava spread out between subsurface layers in horizontal sheets, gradually accumulating over the location of the vent in the shape of a dome, and lifting the overlying rocks into a similar dome-like form. Lava bodies of this type became known as laccoliths.

In addition to the high plateaus, or mesas, there are many other landforms in the Colorado Plateau—ridges, "swells," hogbacks, uplifts, volcanic cones, and more. There are also various kinds of naturally formed holes in the region. Passages cut through the rocks by run-off water are often dramatically deep and tortuous. Long, narrow depressions where, due to faulting, a block of rock has dropped to a level lower than those on either side of it are called grabens. The collapse of underlying soluble material, such as salt, may create steep-sided pits called sinkholes. And various types of holes and fissures occur near volcanic vents. Meteor Crater is not like any of these; but because it lies in a region of unusual landforms, where the effects of past volcanism appear on all sides, it is understandable that many early geologists (few of whom had ever seen it) accepted the opinion reached by G. K. Gilbert, after a brief examination of the great hole, that it had been formed by poorly understood volcanic processes.

Meteor Crater is not the only hole in the Colorado Plateau that was poorly understood in the early years of this century. Less than 200 miles to the east of it, in New Mexico, is another puzzling hole called Zuni Salt Lake crater. The two were frequently compared, and it was felt by some that their resemblances probably indicated similar origins. (In fact, as both of them suggested the kind of landscape that might be found on the moon, both Meteor Crater and Zuni Salt Lake crater were used as locations for the training of astronauts before the first moon landings.)

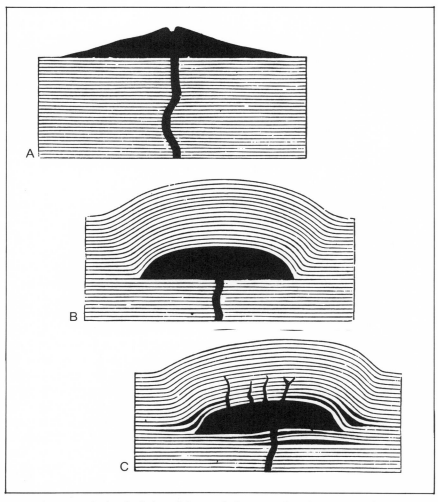

*2.2  Diagrammatic sketch of a laccolith. A: Ideal cross section of a mountain of eruption.*
*B: Ideal cross section of a laccolith, showing typical form and arching of overlying*
*strata. C: Ideal cross section of a laccolith, with accompanying sheets and dikes.*

*2.3    Sinkholes in limestone about thirty miles southeast of Meteor Crater.*

Zuni Salt Lake crater is located about seventy miles south of Gallup. It is roughly circular, a mile or so in diameter, and about 150 feet deep. The shining blue lake which occupies much of its floor contains not clear fresh water but brine, and the pebbles and grasses around its edge are coated with snow-white salt. At one end of the lake two black cone-shaped hills rise steeply, and the taller one reaches a height of almost 150 feet. Otherwise the floor of the depression is fairly level, and parts of it are covered with glistening salt and mud flats. A few buildings erected by various small-scale commercial enterprises which have harvested salt from time to time stand by a road which follows the edge of the lake, and near the two black hills are some evaporating pans. These are the only indications of human activity.

Zuni Salt Lake crater was visited in 1873 by E. E. Howell, a geologist with the Wheeler expedition, who described its basin as anomalous in character and felt that with data then available its origin was inexplicable.[3] Other geologists were similarly puzzled. In the crater rim, besides fragments of rock of the kinds to be found in

**2.4    Zuni Salt Lake, seventy-four miles south of Gallup, New Mexico.**

the surrounding plains, there were thin layers of fine ash, along with grains and pellets of volcanic material known as lapilli, and occasional chunks of lava. All this, as well as the black cone-shaped hills which proved to be composed of fresh basaltic cinders, strongly suggested volcanic action. Nevertheless, many early geologists were of the opinion that no volcanism had been involved, because Zuni Salt Lake crater was quite unlike most known volcanoes. No extensive congealed lava flows and no thick layers of ash gave evidence of great eruptions in the past. Moreover, it was a depression in the land surface, whereas most volcanic craters are elevated above the surrounding countryside by a "volcanic pile" of extruded material. Even volcanic calderas, formed by crustal collapse close to volcanic vents, occur in the higher parts of volcanic mountains.

About the turn of the century it was suggested that Zuni Salt Lake crater might be a sinkhole, caused by solution of underlying salt deposits.[4] This process, however, would create no rim around the resulting hole, and Zuni Salt Lake crater was edged by a slightly upraised rim. It was eventually agreed that if volcanic action had indeed been involved, it was of a little-known kind.

Holes ascribed to volcanic explosions but not connected with outpourings of lava were known in various places. In 1885 I. C. Russell published a geologic history of Lake Lahontan (which once covered parts of what is now Nevada) in which he described two alkaline lakes north of Ragtown, Nevada. They occupied circular depressions which he called "crater rings," the walls of which were composed of layered ash, lapilli, and other volcanic materials. Some-

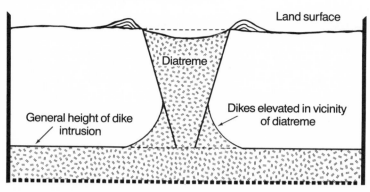

2.5   *Diagrammatic sketch of a diatreme.*

what similar craters of the Eifel district in Germany were called "maars." Many of them, like Zuni Salt Lake crater, were occupied by lakes. Others, containing no lakes, were known as "dry maars." In America, structures apparently similar to dry maars had been called "explosion craters." Two examples, the Ubehebe Craters, may be seen at the north end of Death Valley in California.[5]

Investigation revealed that each maar was the result of brief volcanic activity—a single explosion, or a short series of explosions—involving the release of gas along with small amounts of volcanic material. Vents caused by such volcanic explosions are usually funnel-shaped and filled with rock fragments and volcanic debris. They are known as "explosion pipes" or "diatremes."

In 1942, J. T. Hack published a study of a great many closely spaced and partially eroded diatremes in the Hopi Buttes (Arizona), which were the result of volcanism that occurred several million years ago. Erosion during the intervening millions of years has partly destroyed some of them. Hack found that the explosion pipes flared widely at their upper ends due no doubt to violent expansion of the gases as they reached the surface. The eroded remnants of ancient maars, with layered rims indicating a series of explosions, lay on the flaring upper ends of the diatremes. Where erosion had stripped away the earth below what had once been the ground surface, it could be seen that the diatremes were frequently connected by dikes. Hack observed also that after the volcanic action had apparently ceased, some of the maars had sunk down into the diatremes, the throats of which were filled with rocks and debris.

Theories of the volcanic action of maars were vividly confirmed in 1955, when the creation of a maar was actually witnessed in a volcanic region in the southern Andes in Chile. Beginning in July of that year, violent gaseous discharges occurred, accompanied by smoke and ash. They lasted for twenty or thirty minutes and were separated by similar periods of quiet. Both the explosive periods and the inter-vening pauses grew gradually longer. By November 1955, all activity had ceased. A few months later, water in the new maar (now known as Nilahue Maar) reached its approximately constant level.[6]

Zuni Salt Lake crater is a fairly representative maar, though the presence of cinder cones in the center is unusual.[7] It is thought that at the beginning of its formation a small lava flow emerged from a fault. Before it had completely hardened it was shattered by the explosion of a huge bubble of steam or volcanic gas which probably rose from groundwater in contact with hot rock or molten lava in the fault below it. Both the lava flow and the rocks of the plain on which it lay were shattered by the explosion, and fragments of both fell in a ring around the place from which the gas emerged. Other gaseous explosions followed, each accompanied by small amounts of volcanic debris, which fell in thin layers on the ring of rock fragments around the hole, building it up into a rim. During these explosions the vent was probably widened by slumping of its walls, and the final weak explosions emitted cinders which filled the throat of the explosion pipe. Later, as underlying lava cooled and retreated deeper into the earth, surface debris sank down into the explosion pipe, and sections of its edge slumped into the depression creating the steep sides to be seen today. Lake water was supplied by saline springs which rose along fractures in the rock beneath the maar.[8]

Other maars exist in the western United States. During the past couple of decades interest in their probable origin and peculiar struc-ture has greatly increased and many more, previously unknown, have been identified. Detailed studies have suggested a probable sequence of events during the maar-forming explosions which offers an expla-nation for perplexing sedimentary features often found within the rims. Although no doubt there is still much to be learned about the processes involved in the formation of maars, some general under-standing of them has been achieved, and is the result of work by many people. However, as often happens in such cases, information does not spread very fast. In the early 1970s, some men working at the salt pans in the Zuni Salt Lake crater were asked if they knew how the strange depression had formed. "Sure," one of them replied without hesitation. "It's a meteor crater. Like the one in Arizona."

## CHAPTER 3

# Meteor Crater, Arizona

THE LOCATION OF METEOR CRATER (also known as the Barringer Meteorite Crater) used to be described as near Canyon Diablo, a few miles south of the railway, but today it can be reached by a road which runs south some five miles from the interstate highway, about twenty miles west of Winslow. The crater is a little less than 4,000 feet (1,200 m) in diameter and about 570 feet (about 175 m) deep. Its rim rises an average of about 150 feet (about 45 m) above the surrounding plain. It contains no lake, though flat-lying sediments containing small shells indicate that a lake once existed there.

Meteor Crater was formerly known by a variety of names, the most common of which were Coon Butte, Crater Mountain, and Franklin Hole.[1] It first attracted attention in the 1880s when some sheepherders reported finding silver nearby. The silver proved to be nickel-iron. A few years later a prospector reported that an unbelievably rich vein of iron came to the surface several miles southeast of Canyon Diablo. Pure metallic iron could, he claimed, be picked up from the ground by the carload. Samples of it eventually reached A. E. Foote, of the Foote Mineral Establishment in Philadelphia. The peculiar pitted surfaces of the fragments and the remarkable crystalline structure which could be seen on fractured pieces convinced him that they were of meteoritic origin. In 1891 he journeyed to Crater Mountain to investigate for himself.

Foote described the place where the iron fragments were found as a nearly circular elevation, the center of which was occupied by a cavity with sides so steep that animals had become trapped in it and had left their bleached bones on the bottom. He observed the fragmented rock of the crater rim, and that the horizontal layers of sandstone and limestone near the top of the cavity were uplifted on all sides so that they sloped outward away from the crater at a fairly uniform angle of more than 30 degrees. As he discovered no volcanic rocks he was, as he said, "unable to explain the cause of this remarkable geological phenomenon."[2]

A scouring of the plain showed that carloads of iron existed only in the imagination of the prospector. However, two large masses of iron were discovered, weighing respectively about 150 and 200 pounds. With the assistance of five men, more than a hundred smaller pieces were found. Although Foote's photographic equipment was damaged during a strenuous ride over the plains and his photographs of the geologically puzzling rim were lost, a number of metallic fragments were successfully transported home. Samples were sent to G. A. Koenig, professor of chemistry at the University of Pennsylvania, and analysis showed them to be a nickel-iron alloy, with minute quantities of platinum and other elements. Etched surfaces revealed the definite crystalline structure of Widmanstätten figures. The metal was of extraordinary hardness, and several chisels were ruined in the process of obtaining sections of it. During attempts to polish the surfaces, inconvenient impurities which tore up the emory wheels proved to be extremely small black diamonds—of little commercial value, but of great mineralogical interest. Only four years earlier, diamonds had been found in meteoritic stone. This, apparently, was the first discovery of them in meteoritic iron. (It was later

*3.1   Meteor Crater, Arizona.*[3]

shown, however, that diamonds had been discovered in the iron meteorite Magura by E. Weinschenk in 1889.)

These investigations focused considerable attention on Crater Mountain, or Coon Butte, the more generally used name at that time. The public did not exactly flock to see it, for it was less easily visited then than now; but it was the subject of a number of lectures, of much speculation, and of great interest. No one was more interested than the great American geologist, Grove Karl Gilbert.

Gilbert was a man of great stature in the scientific world. Universally liked and respected, he enjoyed tremendous admiration and prestige. He had been a member of the Geological Survey under J. S. Newberry, of the U.S. Surveys and Explorations West of the 100th Meridian under George M. Wheeler, and of the Geographical and Geological Survey under J. W. Powell; and he remained a prominent and respected member of the U.S. Geological Survey until his death.

In 1891 Gilbert, who was then chief geologist of the U.S. Geological Survey, arrange to have a colleague, W. D. Johnson, make a detailed study of the Arizona crater. In due time, Johnson reported that in spite of the absence of volcanic materials it was his opinion that the crater had been formed by a steam explosion. This conclusion, however, did not explain the presence of meteoritic iron. Gilbert remained curious and arranged to visit Coon Butte himself. "The errand is a peculiar one," he wrote to a friend. "I am going to hunt a star."[4] In November 1891, Gilbert and Marcus Baker, of the U.S. Coast and Geodetic Survey, examined the enigmatic crater. Because it was round, Gilbert assumed that the meteorite, if there had been one, must have fallen from directly above, and must therefore lie buried beneath the hole. But a careful magnetic survey of the crater floor gave no indication of the presence of buried metal. Gilbert calculated that the amount of material in the rim would just about exactly fill the hole. This, too, he felt argued against the presence of a large metallic mass beneath the crater floor; if such a mass were buried there, an equal volume of material must, he thought, exist in the rim. He finally decided that his suspicions as to the meteoritic origin of the crater were baseless after all. "I did not find the star," he wrote to his friend, "because she is not there."[5] In agreement with Johnson, he concluded that the crater had been formed by a steam explosion.

The fact that some question remained in Gilbert's mind can be seen from his contribution to the Fourteenth Annual Report of the U.S. Geological Survey, in which he said that discussion of the two hypotheses for the origin of the crater had led him "to the pursuit of two extensive lines of inquiry corresponding severally to the two hypotheses." He was "thus led to search the literature of volcanoes for the purpose of learning the peculiar characteristics of convulsions wrought by the explosion of steam" and "to give attention to the crateriform hollows of the moon, which have been ascribed by some writers to the impact of meteoric[6] masses falling to its surface."[7] The latter inquiry he viewed chiefly as a private study, but he was permitted to use the instruments and library of the U.S. Naval Observatory. For several months after his return from Arizona, Gilbert spent every available evening before a telescope, studying the moon. This activity drew the censure of at least one congressman who, during a congressional discussion of the shortcomings of the U.S. Geological Survey, declared that "so useless has the Survey become that one of its most distinguished members has no better way to employ his time than to sit up all night gaping at the moon."[8] Gilbert is said to have remarked that certain congressmen and clouds were equally obstructive.

In his laboratory at Columbia University, Gilbert performed experiments (which he called his knitting) in an attempt to gain a better understanding of the impact of one material upon another. He dropped marbles into porridge, and lumps of mud into beds of mud. With an improvised slingshot he fired pellets of clay into beds of clay, and using a large-bore musket he shot balls of various materials into beds of the same material. These experiments seemed to support the idea that lunar craters are the result of meteoritic impact, but the fact that almost all lunar craters are round puzzled him. His experiments seemed to indicate that impact at an oblique angle would create an elliptical crater. To solve this difficulty he invented a theory of "moonlets"—bodies which once formed a Saturn-like ring around the earth, but which gradually coalesced and became the moon. Visible lunar craters were, he suggested, scars of high-angle collision created when the last of the moonlets united with the main body of the moon.[9]

Ever since Galileo's first telescope began to reveal the secrets of our nearest celestial neighbor, astronomers have been perplexed by the craters on the moon. Robert Hooke in 1667 compared them to bubbles in boiling alabaster. Other early investigators maintained that they must have been caused either by volcanic action or by something falling from space, but the question remained unresolved. William Morris Davis once pointed out with wry humor that astronomers tended to explain the craters of the moon by volcanic action, a geologic process, while geologists tended to explain them by meteoritic action, an astronomic process—each scientist evidently feeling free to take liberties with a field other than his own.[10]

In 1892, as retiring president of the Philosophical Society of Washington, Gilbert gave an address, *The Moon's Face: A Study of the Origin of its Features,* in which he pointed out differences which he had observed between lunar and terrestrial craters. He noted that lunar craters were apparently distributed at random, while active volcanoes on the earth were found in regions and belts of volcanic activity. He observed that many lunar craters are of much greater diameter than the largest known volcanic craters on the earth; that they are remarkably round, while terrestrial craters tend to be irregular; and that their floors are usually depressed, while the floors of the earth's volcanic craters are built up by accumulations of lava. Terrestrial volcanoes often create cone-shaped mountains with a small vent at the top; Gilbert found no similar lunar form. Many lunar craters exhibit a central peak; there seemed to be no similar terrestrial form. Gilbert agreed with some observers who claimed that lunar craters resembled terrestrial maars. He was convinced, however, that lunar craters are formed not by volcanic action but by meteoric impact, a conclusion quite opposed to the theories of volcanic processes favored by most scientists at that time.

In 1895 Gilbert gave another presidential address, this time before the Geological Society of Washington. His title was *The Origin of Hypotheses, Illustrated by the Discussion of a Topographic Problem.* The topographic problem was that of Coon Butte, the Arizona crater. Gilbert described the scientific approach to problems in general as being the developing, followed by the careful testing, of hypotheses; the hypotheses might be intuitive or suggested by analogy. This was not the first time he had set forth these principles, to which he was utterly dedicated. Similar ideas had appeared in a paper written in 1886 in which he remarked, "In the testing of hypotheses lies the prime difference between the investigator and the theorist. The one seeks diligently for the facts which may overthrow his tentative theory. The other closes his eyes to these, and searches only for those which will sustain it."[11]

To illustrate this scientific method, Gilbert related how, in his examination of Coon Butte, various hypotheses concerning its origin had been tested and discarded. The only hypothesis which survived was that of formation by a steam explosion. Everything discovered in the crater was then considered with reference to this last hypothesis, and as Gilbert remarked, although all were not yet understood, they seemed not to oppose the theory. He noted that "the little limestone crater" is in the midst of a great volcanic district, and thought that it might prove to be similar to maars in Germany and France. He was troubled, nevertheless, by the highly improbable coincidence of meteoritic iron having fallen at the precise spot where a steam explosion occurred. In closing, Gilbert mentioned a question asked by a correspondent which, he said, had the "ability to unsettle a conclusion which was beginning to feel itself secure." The correspondent had asked whether some of the rocks beneath a buried star might be compressed by the shock of impact so that they occupied less space—which, if true, would invalidate part of the argument on which the accepted hypothesis was based. Such an unsettling question, Gilbert remarked, illustrated the tentative nature of what science calls results, which "are ever subject to the limitations imposed by imperfect observation. . . . In the domain of the world's knowledge there is no infallibility."[12]

In spite of what amounted (at least in this address) almost to a suspension of judgment on Gilbert's part concerning the origin of Coon Butte, the effect of his words was far from what he intended. What most people heard, rather than admonitions for scrupulous care in the search for facts, was that Grove Karl Gilbert said that Coon Butte had been created by a steam explosion.

A mistake, like the evil that men do, lives after them. Good scientific work, while not exactly interred with their bones, is often so completely assimilated that it becomes an anonymous part of the discipline they have helped to develop. However, this was not entirely true in Gilbert's case. Because of his interpretations of various previously unexplored regions, his immensely influential work on erosion and fluvial mechanics, and the principles he defined concerning the development of landforms, he is still a towering figure in geology. His ideas and observations about lunar craters, which in his own day were rather startling and were possibly obscured by his "moonlet" embroidery, are now considered to have been far in advance of his time. But in the recognition and study of meteoritic craters, a delay of several decades is attributed by many people to Gilbert's mistake.

During the first quarter of this century, geographical and geomorphological thought in America was dominated by William Morris Davis, whose "cycle of erosion" interpreted the development of landscapes through successive periods of youth, maturity, and old age. According to Davis, any landform could be understood by considering its basic structure, the natural processes affecting it, and the stage of development which it had reached. Davis was a prolific and persuasive writer, and his ideas were widely accepted and extremely influential.

Some twenty-seven years after Gilbert's unfortunate pronouncement, Davis published a biographical memoir of Grove Karl Gilbert. He recalled that during 1895 Gilbert spoke at several universities on the subject of Coon Butte. The popularity of this lecture, Davis remarked, "was due much more to the philosophical manner in which its subject was treated than to the inherent importance of the little geographical feature on which it was based."[13] As to the origin of Coon Butte, Davis dismissed the matter rather lightly. He remarked, without taking a position one way or the other, that the presence of meteoritic iron in the immediate vicinity argued in favor of meteoritic impact, while the presence in the immediate neighborhood of other craters of unquestionably volcanic origin argued in favor of explosion. But, Davis went on, Coon Butte was after all a "relatively trifling and local affair,"[14] and the important part of Gilbert's speech had been his emphasis on the invention and testing of hypotheses.

Actually, questions as to the origin of such a crater were quite outside the field of Davis's real interests. During the Transcontinental Excursion of 1912, a tremendous undertaking by the American Geographical Society, involving a special train and distinguished participants from home and abroad (and organized largely by Davis), a visit was made to the Arizona crater and supper was served to the whole party on its rim.[15] Davis seems to have taken part in these proceedings without suffering any particular assault upon his imagination, and such was his prestige at that time that if he felt a matter to be unimportant few would venture to disagree with him. But a few there had been, over the years. One of them was a geologist and mining engineer, D. M. Barringer.

During the 1890s, as we have seen, several people became interested in the Arizona crater—because it existed, because of its diamond-bearing meteoritic iron, and because of speculations about its origin. Nevertheless, dissemination of information about it was spotty, to say the least. It was not until 1902 that Barringer chanced to hear of it. Though incredulous, he was interested in reports that, official opinion notwithstanding, many local people still considered the great hole to have been formed by a large body of iron that fell from the sky.

Barringer mentioned the matter to a friend, Benjamin C. Tilghman, a mathematician and physicist, who was an authority on impact projectiles from heavy ordnance. He found Tilghman's interest as keen as his own. After examining a number of specimens of Canyon Diablo iron, both men were convinced, even before visiting the crater, that the local people were correct: the great hole must have been created by the fall of a meteorite. They felt sure that a large amount of meteoritic metal lay beneath the crater and, forming a small mining company (the Standard Iron Company), they acquired the mineral rights necessary before beginning serious efforts to locate the meteorite.

They examined the crater in great detail—by 1905 Barringer had visited it more than ten times—and together they collected what he described as an astounding array of evidence in support of their opinion. They confirmed Foote's statement that the crater contained no volcanic rocks (except for some thin layers of volcanic ash in the lake sediments at the bottom, part of a dusting of ash spread over the whole region by an eruption at the nearby Sunset Crater, now thought to have taken place in the decade between A.D. 1060 and 1070). They also observed, as Foote had done, that the uppermost layers of rock exposed in the walls of the crater were raised considera-

bly above their level in the surrounding plain and that they slanted
outward and downward from the crater. There were even some sections which appeared to have been thrown out bodily and which lay
overturned on the rim with their successive geologic layers in reverse
order.

They observed that fragments of the three upper layers of regional
rock (from depths down to about 1,000 feet) were scattered concentrically around the crater out to a distance of about two and a half miles.
More than a mile from the rim Barringer reported finding enormous
boulders of limestone and sandstone which he estimated to weigh
some 5,000 tons. It did not seem possible to him that a steam explosion could have exerted the force necessary to hurl such immense
boulders over such great distances, and in any case stones hurled
from explosion craters are usually composed of volcanic materials
and come from depths greater than that of the upper rock layers. In
addition, if there had been sufficient heat and underground water to
create a steam explosion of such magnitude, there would probably
have been a series of events. Nothing, however, was found to suggest
more than a single outburst. There was no evidence of hot spring
activity, and the rim showed no indication of layering. In the rim, on
the contrary, boulders and rock fragments and pieces of meteoritic
iron and fine debris were indiscriminately mixed.

Vast quantities of powdered rock (which they called "silica") were
found in both the rim and the crater. Under the microscope this
powdered rock, in some cases so fine that no grit could be discerned
when it was placed between the teeth, was seen to be made up of
broken grains of quartz sand which looked like angular pieces of ice.
To both men it seemed impossible that such pulverization of the sand
grains composing the rock could have been caused by volcanic action
of any kind. They estimated that hundreds of millions of tons of
"silica" lay in the crater and rim, and interpreted it as due to the
pulverizing effect of an almost inconceivably great blow. They also
found large quantities of magnetic iron oxide, or "iron shale," different from any natural substance with which they were familiar, the
larger pieces of which on analysis almost always proved to contain
nickel.

Tilghman claimed that Gilbert's method of magnetic detection
would probably have discovered a large mass of solid iron, had there
been one, but would not have revealed a mass of fragments. Therefore, if the meteorite had arrived in a fragmented condition, or if it
had shattered on impact, its presence could not have been detected
with Gilbert's instruments. Tilghman also differed sharply with Gilbert's estimate of the amount of material in the rim. According to
Tilghman's calculations, the rim contained considerably less material than would be necessary to fill the hole. He felt this to be an
irrelevant matter, however, as a strong wind at the time of impact

*3.2   Boulders and rock fragments on the rim of Meteor Crater.*

might have blown away enormous quantities of dust, and an unknown amount of material had no doubt been removed by erosion
during the intervening time. He suggested also that a meteorite
traveling at great speed through the atmosphere must have been
"surrounded by the usual tail always accompanying such objects."
The luminous tail no doubt consisted of very small metallic particles.
With the use of magnets a search was made for such material, and its
presence, in the form of a "blackish gray, rather fine-grained powder,
strongly attractable by the magnet," was found to be "absolutely
universal over the whole locality."[16]

As part of the work of exploration, several men were employed to
search for "irons." Although tons of metallic fragments had already
been removed by collectors, many more pieces of metal were found,
and their locations were plotted on a chart. It became clear that they
were concentrated most plentifully in a crescent-shaped area, the
inner edge of which coincided precisely with the edge of the crater.
Only two or three metallic fragments were found within the crater, in
locations suggesting that they had been washed down from the rim.
All this seemed to provide ample proof that the irons had been emplaced at the moment of impact: if there had been a separate fall
earlier, they would have lain beneath the ejected debris, instead of
being mixed through it; and if later, they would be found scattered
within the crater as well as on the rim.

In an effort to locate the meteoritic mass, borings were made
through the lake sediments of the crater floor, revealing hundreds of
feet of powdered rock, or "silica," below which were zones of crushed

rock, and finally, undisturbed sandstone. There was no evidence of volcanic action, nor was there any evidence of a huge mass of metal.

Accounts of their findings were published in separate papers by Barringer and Tilghman in 1905. The great interest they generated was unfortunately concentrated in a rather few individuals. Many people regarded their work as that of a couple of unknowns brash enough to oppose the word of a scientist of great repute, and it seems possible that at the time the papers may have remained largely unread. A few people, however, read them with care and considerable excitement. One was J. C. Branner, of Stanford University, who promptly made a trip to the crater. Various features of the place were pointed out to him by S. J. Holsinger, a member of the mining company, who had spent four years camping on the rim while exploration was in progress. Branner was soon convinced of the meteoritic origin of the great hole and later stated his views at a meeting of the American Association for the Advancement of Science. Another interested person was H. L. Fairchild, of the University of Rochester, who also visited the spot. He agreed with Branner that if the crater was not of meteoritic origin, it was the most interesting geological puzzle of the present time. He presented arguments for the impact origin of the crater at the Tenth Session of the International Geological Congress in September 1906. Fairchild had been the secretary of the Geological Society of America for sixteen years. On his final appearance in that capacity, at the annual meeting of the Society in December 1906, he laid before its members the new facts supporting the impact origin of the Arizona crater and formally proposed the name Meteor Crater.

Perhaps the most influential geologist to be favorably impressed by the work of Barringer and Tilghman was George P. Merrill, curator of the Department of Geology of the United States National Museum, professor of geology at what is now George Washington University, and the author of several books on geological subjects. He was particularly interested in stones and was one of the first to apply petrological methods to the study of meteorites. Merrill suggested that with so much new information which had not been available to Gilbert another investigation of the crater was in order. This suggestion was approved by the secretary of the Smithsonian Institution. Gilbert, who had not changed his opinion and who was at that time in very poor health, placed his maps and charts and other records at Merrill's disposal.

Merrill spent several days at the crater in the company of Barringer and Tilghman, with whom he soon found himself—possibly somewhat to his own surprise—in rather general agreement. He made a scrupulously careful geological and mineralogical study of the crater and agreed that the crushed and shattered rock in the walls, the peculiar metamorphism observed in fractured and distorted quartz grains, and the enormous quantities of "silica," which he preferred to call "rock flour," must be the result of a tremendous concussion. Borings revealed an abundant slaggy material, a peculiar pumice-like form of sandstone, beneath the crater floor. Tilghman discovered that it contained both iron and nickel, and it was noted by H. L. Fairchild in a paper read before the Geological Society of America as "Variety B." (For Variety A, see chapter 10.) Microscopic examination revealed that the fused sandstone was a quartz glass. It was described by Merrill as closely resembling fulgurite glass, which is sometimes formed when lightning strikes in sand and which requires for its formation a temperature higher than any recorded in volcanic eruptions. He concluded:

> As to the exciting cause of this metamorphism, so far as the writer has information, no more satisfactory theory has been advanced than that of the Messrs. Barringer and Tilghman, who ascribe it to the impact and incidental heat of an enormous mass of meteoric iron which constituted a portion of the well-known Canyon Diablo fall. Startling as it may seem, the writer, without intending to commit himself in any way, has to acknowledge that it must at least receive consideration, for the simple reason that nothing else seemingly worthy of consideration presents itself.[17]

Merrill's caution reflected the skepticism with which the possibility of a terrestrial meteorite crater was generally viewed at that time. Though it seemed to him probable that vaporization of considerable meteoritic material—possibly almost the entire meteorite—had occurred at the moment of impact, he found little in the way of vaporization products; nonetheless, the fused quartz sandstone provided tangible proof that temperatures at least as high as 3,900° F had existed. He remarked:

> It is well-nigh impossible, however, that a force so great, and applied, as apparent, in an instant of time, should not have been productive of an amount of heat so vastly greater than 3900°[F] that its expression in figures would be utterly meaningless and incomprehensible, and in the writer's mind the greatest difficulty in accepting the meteoric hypothesis lies in the absence of sufficient evidences of such extreme temperatures. There are no volatilization products and but slight evidence of slags among the products thus far brought to light. Only the fused quartz remains as tangible proof.[18]

Merrill's suggestion of volatilization of the meteorite was not accepted by Barringer, who believed that if such an event had taken place vaporized metal would have left revealing stains "on a grand scale" on the rocks, and no such stains were to be found.

In spite of the efforts of Branner and Fairchild and several others, and in spite of Merrill's support for the meteoritic hypothesis of Barringer and Tilghman for the origin of the Arizona crater, Gilbert's words still prevailed in many quarters. The origin of the great hole remained a matter of uncertainty.

By July 1908, some twenty-eight holes had been drilled, at enormous expense, into the floor of the crater. Small metallic fragments which dulled the drills and quicksand composed of water-saturated "silica" complicated the work and usually necessitated abandoning the locations involved. But there was no indication of a mass of meteoritic metal. Tilghman, who had invested many thousands of dollars in this work, lost confidence in its commercial possibilities and dropped out in 1909. Barringer, too, was deeply committed financially but remained convinced that a metallic mass must lie buried somewhere in the crater region; he was encouraged by interest and supporting opinion that continued to come from various authorities. One of them was H. N. Russell, head of the astronomy department at Princeton University, who stated after an examination of the crater that he was thoroughly convinced of its meteoritic origin.[19] Another was Elihu Thomson, director of the Thomson Laboratory of the General Electric Company, who remarked in a letter to Barringer that "this Arizona crater bears all the evidences of impact and the evidences of nothing else."[20] W. F. Magie, professor of physics at Princeton University, spent three weeks at the crater. In his opinion, there was no doubt that the great hole was formed by a meteor "and this meteor is buried in it."[21] In 1910 he conducted a magnetic survey of the crater, and because he found no indication of a large metallic mass he concluded, in agreement with Barringer and Tilghman, that the meteor probably consisted of a cluster of small masses of iron. He also pointed out a remarkable symmetry in the distribution of ejected material, noting that it was symmetrically placed on either side of an axis that ran from a little west of north through the center of the crater. (Bilateral symmetry was later recognized as a common feature of meteorite craters.) The work of Barringer and Tilghman was known far beyond the limits of the United States, for it was suggested at a meeting of the Geological Society of Stockholm in 1910 that the Lake Mien structure in southern Sweden was of impact origin,[22] and it was compared with the Arizona crater.

"A significant fact," Barringer wrote in 1914, "is the discovery . . . in cores obtained by drilling . . . of quite large quantities of quartz glass, which is undoubtedly fused sandstone and has been so described by Merrill. . . ." He continued:

I am assured by Dr. Merrill and others that there is no record of a sudden outburst of volcanic action wherein the heat generated was sufficient to fuse crystalline quartz. . . . It seems to me therefore that this peculiar vesicular form of metamorphosed sandstone, which was certainly produced by a sudden and intense heat and which is so abundantly stained with nickel and iron oxide, in itself furnishes an incontrovertible proof of the impact theory of origin, the opinion of certain members of the U.S. Geological Survey to the contrary notwithstanding."[23]

Even this statement, however, did not immediately counterbalance the effect of Gilbert's steam explosion theory. To many people at that time, including many scientists, the idea of a meteorite crater on the surface of the earth still suggested the fantastic. And while it was rumored that Gilbert had perhaps changed his mind about the matter, no announcement was made to that effect.

During the early exploratory work, drilling had been done in the approximate center of the crater floor, on the theory that because the crater was round, the meteorite must have fallen from directly above. After experimenting with rifle bullets fired into rock, Barringer noted that a round hole was created even when the projectile came from a direction several degrees from the perpendicular. He suggested that the meteorite might have struck obliquely and lodged beneath the south rim, where an upward bulge could be discerned. As funds were too low to attempt further drilling, Barringer spent much time trying to convince skeptics that the crater was meteoritic and to attract capital in order to continue attempts to find the meteorite. This was a frustrating and disheartening task, carried on through what his sons described as "his bulldog persistence in the face of apathy, opposition, and even ridicule."[24]

At last, in the early 1920s, satisfactory arrangements for further work were made with a mining company. An exploratory boring was made in the south rim of the crater, but work came to an abrupt halt when at a depth of 1,376 feet the drill became wedged and was lost. Barringer optimistically interpreted this mishap as contact at last with the main mass of the meteorite. New borings were started with the intention of making subterranean connections with the wedged drill, but when only about half the necessary depth had been reached, a strong flow of underground water made further progress impossible.

A year or two later money was raised for still another attempt. It, too, was extremely expensive—and unsuccessful. By the time of Barringer's death in 1929, more than $600,000—a tremendous sum in those days—had been spent on exploration of the crater. No meteorite had been discovered. The meteoritic origin of the crater, however, was at last finding acceptance among scientists. In fact, new discoveries were drawing increasing attention to the whole little-understood subject of meteorite craters, and independent investigations were being carried out which confirmed the statements of Barringer and Merrill. For example, A. F. Rogers, who had studied samples of metamorphosed sandstone, or silica glass, from Meteor Crater, concluded as both Barringer and Tilghman had done that the silica glass (lechatelierite) was produced by fusion of the Coconino sandstone. A temperature higher than any known to have occurred in volcanic processes must therefore have existed at the time of the fusion. "The impact of an iron meteorite with a diameter of 500 feet or so," Rogers concluded in 1930, "would probably produce a cavity of the size of Meteor Crater, and there would probably result, from friction, a temperature sufficiently high to melt quartz."[25]

Over a period of some twenty years, a tremendous amount of capital was invested in the Arizona crater. Much was learned, but nothing indicating that the enterprise was about to become profitable. By the late 1920s, Merrill's unwelcome suggestion that the meteorite might have been destroyed in its collision with the earth was finding support among scientists. It was felt that before investing further capital, definite information should be obtained as to the depth and location of the buried meteorite. Shortly before Barringer's death, advice was sought from F. R. Moulton, a well-known mathematician and astronomer who for many years was a professor at the University of Chicago. He had been noted for ballistic research during World War I, and was known for his part in development of the Moulton-Chamberlin planetesimal hypothesis, which suggested that planets were formed by the aggregation of smaller bodies.

For Moulton, the problem presented by the Arizona crater was an unfamiliar one. Among his initial responses was a suggestion that the meteoritic mass might be located by the use of geophysical methods, provided that it "is as large as seems probable, and does not lie at a depth greater than 5,000 feet."[26] But Moulton continued to study the matter, and his report, made in August 1929, was dismaying. In his opinion, the meteorite was probably much smaller than earlier estimates (of five to fifteen million tons) had suggested. Moulton thought it was probably no greater than three million tons, and possibly as small as fifty thousand, and that rather than being concentrated in an easily minable mass, it was probably scattered through a hundred times its weight of earthy material. This caused so much concern among the people connected with the mining company that Moulton shortly produced a second, more detailed report. This one was even gloomier. He noted that in his first report he had made "liberal allowances on every hand so as to obtain a safe upper limit to the mass." In the second report he was "trying to come closer to the facts."[27] and he calculated that the mass of the meteorite was of the order of only a hundred thousand tons, and that it was probably destroyed on impact.

Moulton was well aware of the blow he was delivering, for the unmistakable conclusion to be drawn from his reports was that further mining was inadvisable. However, he did not expect his advice to be taken. Should further investigations be made (which he thought likely), he suggested that if more than five hundred thousand tons of meteorite were found (which he thought unlikely) everyone should "take a long and hearty laugh at my expense, and call on me to put up a dinner at the crater for all interested."[28] In 1931 in his textbook on astronomy Moulton remarked that

> the energy given up in a tenth of a second would be sufficient to vaporize both the meteorite and the material it encountered—there would be in effect a violent explosion that would produce a circular crater, regardless of the direction of impact, which alone would remain as evidence of the event.[29]

Although Moulton's reports were bitterly disappointing to the mining interests, greater understanding of what takes place during a collision of a meteorite with the earth provided an explanation for lack of success in the search for a buried meteorite. It was not surprising, however, that men who had for many years faced skepticism concerning their ideas about the origin of the Arizona crater, and who had invested time and money in mining it, should be unwilling to accept Moulton's verdict. They had a geophysical survey made of the crater. "Material of metallic character" was indicated, and borings were made to reach it.[30] Some material that gave a positive test for nickel was found, but very little meteoritic metal. Moulton was not called upon to produce a dinner.

## CHAPTER 4

# More Meteorite Craters

BY THE LATE 1920s BARRINGER'S work had convinced many scientists that the great hole in the Arizona plains was a meteorite crater. And if one meteorite crater existed on the surface of the earth, no doubt there were others. This soon proved to be the case.

In 1926 a letter from A. B. Bibbins, of Baltimore, appeared in a mining journal. Bibbins explained that growing interest in meteorites had reminded him of an experience in Texas where, in 1921, in the course of a search for potash and oil near Odessa, a ranchman had given him a sample of what was thought to be iron ore. He was taken to the edge of a "blow-out" (a depression caused by wind erosion, common in parts of the western plains) where the sample had been found, and where after a short search a similar one was discovered. The depression was roughly circular, a few hundred feet in diameter, ten to thirty feet deep, and seemed to be fringed with rock debris. Pressure of time prevented a detailed investigation, but Bibbins suspected that the iron was meteoritic, and it was sent to the U.S. National Museum, where analysis confirmed his suspicions. A description of it was published by G. P. Merrill in 1922.

Bibbins's letter chanced to catch the eye of D. M. Barringer, who, with some excitement, pointed it out to his son. D. M. Barringer, Jr., visited Odessa and located the crater. He discovered that, although badly eroded, it bore significant resemblances not to other wind-eroded depressions—but to Meteor Crater. A relatively low rim containing rock fragments enclosed a shallow depression almost filled with sand. The depression was steep-sided within and gently sloping on the outside, and in some places uplifted rock layers could be seen slanting away from the center to the level of the plain. Several pieces of nickel-iron were found, as well as "iron shale" similar to that found at the Arizona crater. A careful search revealed no evidence of volcanism or of hot spring action. Barringer, Jr., concluded that the depression had been formed by "the impact of an extra-terrestrial body with the earth"—that it was a meteorite crater; and using a pre-arranged code, he telegraphed this information to his father. In a letter written the same evening, he said, "It is a meteor crater beyond the shadow of a doubt. . . . The whole thing is so absurdly like Meteor Crater that it doesn't seem possible. . . ."[1]

The depression became known as the Odessa crater, and it was found to be the largest of a group of small craters. Surface indications of the others had been obliterated by wind-blown sands. Similarities between meteorite fragments from Odessa and Meteor craters suggested that they might be part of the same fall, but chemical analyses later showed that this was not the case.

The Odessa crater was thought to be very recent until the fossilized remains of a primitive horse were discovered buried within it.[2] In 1982 fragmentary elephant remains were also found within the crater.[3] The age suggested at present for the Odessa crater is approximately 10,000 years.

Another report of meteorite craters came from the Baltic island of Saaremaa, or Oesel (Estonia), where several small craters had been known for more than a century. They had been variously described as man-made earthworks, sinkholes due to underlying limestone or salt deposits, expansion of anhydrite, the results of gas explosions, or maars similar to those of the Eifel district in Germany. In 1922 they were compared with the Arizona crater, and in 1927 and 1929 the results of investigations of seven of the craters were published by I. A. Reinvaldt, the Estonian Inspector of Mines.

The largest crater, the Kaalijärv, was occupied by a lake and edged by a tree-covered rim. Its inner slope revealed rock layers slanting radially outward and downward, and beneath the center of the crater was a zone of pulverized and fragmented rock. Reinvaldt noted the bowl shape of some of the craters and suggested that they had been formed by explosions accompanying the fall of comparatively small meteorites. He attributed the explosions to the sudden creation of steam caused by heat generated at the moment of impact. Pointing out that the smaller craters were funnel-shaped rather than bowl-shaped, he noted that they were probably caused by impact rather than by explosion. He concluded that they had been formed by a shower of iron meteorites and that further exploration might reveal more craters, and he attributed the absence of meteoritic iron to the fact that the region had been heavily populated for centuries. In 1937 a number of small fragments of nickel-iron were found, and the meteoritic origin of the craters was confirmed.

In 1927 various newspapers reported that L. Kulik, curator of the meteorite department of the Mineralogical Museum of Leningrad, had led an expedition to a remote part of Siberia where some nineteen years earlier a tremendous explosion had occurred. G. P. Merrill's account of Kulik's expedition, in the form of a "rather free translation," appeared in *Science* in May 1928.

It seems that in the early summer of 1908 a great fiery mass lit up the sky over northern Russia. Its fall into the taiga (the subarctic evergreen forest) was immediately followed by a column of fire which rose into the sky along with clouds of heavy black smoke, a deafening roar which was heard for hundreds of miles, and a terrific air wave said to have knocked down both animals and people. Attempts in 1908 to locate the meteorite were unsuccessful, and it was later determined that the correct locality was hundreds of miles farther north than the searchers had supposed because the intensity of the light had made the distance seem deceptively small. Panic among the inhabitants gradually subsided. Time went by. Interest slowly waned, and the matter was forgotten except as a story among the local people.

Early accounts of this event were confused and slow to emerge. No doubt the turmoil of the First World War and the Russian Revolution gave little incentive for the investigation of such things. It was not until 1921 that Kulik unearthed reports to the effect that at 7:30 A.M. on June 30, 1908, in mild weather and beneath a cloudless sky, a fire-ball had rushed through the air over the basin of the Yenisei River in central Siberia. Following its fall, explosions were heard; earthquakes were recorded by seismographs in the Physical Observatory at Irkutsk, where calculations were made indicating that the epicenter of the "earthquake" was in the region of the Podkamennaya Tunguska (Stony Tunguska River); and during the weeks following, a succession of abnormally light nights was experienced in many parts of Europe.

Funds and equipment for a search were finally procured, and in March 1927 Kulik set out from Leningrad. The comfort of railway carriages soon had to be replaced by sleds. The expedition traveled along reindeer paths through the great pine forests, battling frequent snowstorms and bitter temperatures. As they drew near to the Stony Tunguska, occasional burned patches were observed in the forest. At the factory of Vanovara (a base for hunters) on the Tunguska River, a Tungus family with ten reindeer was hired to act as guides and to handle transportation of supplies. Progress was hampered by deep snow and also by the habits of the Tungus, who had perhaps overadjusted to the brief day of the arctic winter. Travel did not commence until ten in the morning and ceased about three in the afternoon, and much time was spent in drinking tea and looking for lost reindeer. Kulik noted that three or four miles was all that could be achieved in a day.

The reindeer path eventually faded away. However, by continuing to press forward on snowshoes, Kulik reached a hill from which could be seen an immense open region surrounded by bald, snow-covered hills. Trees which must have been blown down by the great air blast lay beside one another on the ground, with their tops pointing away from the center. This, evidently, was the region of the great explosion.

Melting snow made further exploration impossible. Kulik returned to Vanovara, determined to reach the locality again by water. He had rafts built, and with great difficulty supplies and when possible the men themselves were transported on the swollen, ice-laden streams. Rapids presented serious problems; baggage was often carried through white water on the backs of the travelers. When at last the deforested region was again reached, Kulik managed to inspect the great area on foot. Everywhere, for a distance of more than twenty miles from the center, he found uprooted trees which pointed radially outward. Over much of the devastated area there was also evidence of fire. In some of the ravines and deep valleys trees had remained upright but were scorched and lifeless. In the central part of the fire zone Kulik found a number of funnel-shaped craters, the bottoms of which were covered with sphagnum moss, but no meteorite fragments were found. Excavation of the craters was not attempted, for by this time supplies were seriously depleted and the party was nearing exhaustion.

Because the fallen forest lay near the southern limit of permanently frozen ground, later interpretation ascribed the craters discovered by Kulik to the action of permafrost. It was concluded that in a region of semipermanent frost and deep swamp, a meteorite crater which was by then some twenty years old would probably have almost disappeared. So, also, would any meteorite fragments which might have been scattered nearby. Nevertheless, the felled forest provided evidence of what had become known as the Tunguska meteorite, and in a later expedition (1929) Kulik reported finding microscopic globules of nickeliferous iron and fused aggregates of quartz grains in two of the craters. However, some aspects of the Tunguska event remained controversial. There is a widely held opinion that it was caused not by a meteorite but by a comet. (The nucleus of a comet is thought to be composed of various ices—probably those of carbon dioxide, ammonia, and water—along with small amounts of stony or metallic material.)

In 1928, accounts of the great aerial disturbance experienced in Siberia twenty years earlier led to the interpretation of air waves which had been recorded at that time in six different locations in

**4.1** *The South Swamp, thought to be the place where the Tunguska meteorite fell and exploded.*

England—records which previously were not understood. Also, a puzzling phenomenon called "night dawns," when twilight was prolonged until daybreak, had been recorded in many places immediately after the Tunguska event. This was now attributed to a heavy blanket of dust which accumulated in the upper layers of the atmosphere after the explosion.

As F. J. W. Whipple remarked in 1930, "There are many marvelous features of the story of the Siberian meteorite, a story without parallel in historic times. It is most remarkable that such an event should occur in our generation, and yet be so nearly ignored."[4] He went on to point out that if it had occurred twenty years earlier, nothing would have been known of the earth waves it created—for the first long-range detection of earth waves was achieved in 1889, when an earthquake in Japan was recorded on a seismograph in Potsdam, Germany, and a global network of seismograph stations was not begun until the 1890s. And if it had taken place only five years earlier, before the establishment of techniques for detecting atmospheric disturbances (such as the microbarographs in use in some places at that time), there would have been no evidence of the spreading air waves beyond the immediate locality of the fall.

A group of meteorite craters was discovered in Australia as a direct result of interest aroused by a quite unconnected meteoritic event. About an hour before midnight on the evening of November 25, 1930, a large region of South Australia was suddenly lit up by a brilliant fire-ball the size of the full moon. For about six seconds it provided illumination as bright as daylight over a very large area. Showers of sparks issued from it as it approached the earth, and its disappearance was followed by a noise variously described as similar to thunder, to the rumble of an earthquake, to the roar of a passing train, or to the banging together of sheets of galvanized iron.

A few days later, a search party led by Professor Kerr Grant of the geology department of the University of Adelaide set out to find the meteorite. The approximate location of the fall had been determined as close to the township of Karoonda. Questioning of eyewitnesses in this neighborhood resulted, on the third day of the search, in the discovery of a stony meteorite. It lay in a shallow crater-like hole some eighteen inches across in a sandy field with numerous meteorite fragments scattered around it. Its total weight was estimated at about 92 pounds, and it became known as the Karoonda meteorite.

No doubt because of the interest created by this event, Grant was informed soon after, by two different persons living in central Australia, that several crater-like depressions existed near the Henbury Cattle Station, and that meteoritic iron had been found near them. This information was passed on to the South Australia Museum. A preliminary survey of the area in question was undertaken in 1931 by A. R. Alderman, also a geologist at the University of Adelaide.

*4.2 Forest devastation in the region of the Tunguska disturbance.*

**4.3**   *The glow and shining clouds seen in northern Siberia on the night of June 30 – July 1, 1908.*

Henbury, a remote spot in Central Australia, lay some 120 miles (by automobile) from the nearest railway station, a distance shortened somewhat if the journey were made by camel, the usual means of transportation at that time. Here an inconspicuous group of craters was found. Eroded and sediment-filled, they were not easily distinguished from the rest of the arid landscape—except for the fact that, as they apparently acted as catchment basins, trees which usually grew in water courses flourished within them. In an area of about a square half mile, twelve probable craters of various sizes were discovered. Because of the close proximity of the two largest, the spot was known locally as the Double Punchbowl. Fragments of nickel-iron and "iron shale" were scattered about, and near the largest of the craters bits of black glassy material were found, which proved to be heat-fused quartz. The craters were interpreted as having been created by a shower of iron meteorites.

A puzzling crater in Persia (now Iran) was pointed out in 1916 to Brigadier General R. E. H. Dyer, of the British army, by one of his assistants, Idu Kahn Reki. Idu explained that one day when his grandfather was a young man something exploded in the sky and fell to earth, creating a great hole in an otherwise flat plain. He claimed that the hole was originally twice its present depth, but due to erosion the sides had gradually fallen in, causing it to become shallower. Dyer estimated the hole to be 150 feet (45 m) long, 120 feet (37 m) wide, and about 50 feet (15 m) deep, with "smooth, level lips"; and although no meteoritic material was discovered, he concluded that an enormous meteorite must lie buried beneath the bottom.

Measuring it in 1921, C. P. Skrine found it to be an oval shape, about 95 feet (20 m) along one axis and 75 feet (23 m) along the other, and some 35 feet (10 m) deep. He observed that flood waters had created a narrow channel in the wall and that the bottom was evidently silting up fairly rapidly. This impression was confirmed in 1929, when he found the floodwater channel enlarged and the crater at least 3 feet shallower.[5]

Another puzzling crater was known in Ashanti, on the Gold Coast (now Ghana) of Africa. It was described in 1931 by Malcolm Maclaren as roughly circular, with a diameter of some six and a half miles (about ten kilometers) and a continuous unbroken rim rising 600 to 900 feet (185 to 275 m) above the surrounding upland region. Its steep inner slopes were covered with thick jungle, and its floor was occupied by Lake Bosumtwi, a fairly deep body of water with a diameter of about five miles—and the only lake in the region. Maclaren found no volcanic materials which would have supported the commonly held opinion that the crater was volcanic. He suggested that it might have been created by the impact of a stony meteorite which, he felt, could explain the crater and also the absence of meteoritic materials in the vicinity.

In 1932 H. St. John B. Philby, a British explorer connected with the Royal Geographical Society of London, set out to investigate what was known as the Empty Quarter of Arabia, the Rub' al Khali. Philby had spent considerable time in Arabia and was particularly interested in legends concerning the ancient city of Wabar, said to have been destroyed long ago by fire from heaven. The city had been, perhaps, the center of some long-forgotten civilization, and ruins of it were said still to exist among the sands of the Rub' al Khali. Philby traveled slow, difficult miles on long-suffering camels, accompanied by Arab guides who had little enthusiasm for the undertaking. The location of the legendary city was said to be marked by a lump of iron the size of a camel which might, Philby thought, be the remains of some ancient statue.

When eventually the place was reached, they came upon several low, roughly circular structures of slaggy material, almost obliterated by drifting sands, which were thought by the Arab guides to be the blackened walls of devastated buildings. To Philby's disappointment they appeared to be the craters of ancient volcanoes, although lumps of slag which surely must be burned bricks of destroyed houses were

**4.4**   *Henbury craters, about 115 miles southwest of Alice Springs, in central Australia.*

pointed out to him in efforts to convince him otherwise. Small round bits of shiny black material were plentiful and were collected by the Arabs, who thought them remnants of the fabled wealth of the ruined city and optimistically called them "black pearls"—though their market value after the long journey to Mecca proved to be disappointingly low. No camel-sized lump of iron was found, but a smaller piece, rabbit-sized according to Philby, was discovered under some desert scrub. It was rusted and very heavy, and Philby assumed that it was a meteorite.

The previous year (1931), Bertram Thomas, also connected with the Geographical Society of London, had penetrated parts of the Rub' al Khali. Philby carried with him the issue of the *Geographical Journal* which contained an account of Thomas's journey and which he frequently found occasion to consult. By an interesting coincidence this issue also contained Maclaren's article on the "supposedly meteoritic crater at Bosumtwi, in Ashanti," and Philby was led to speculate as to whether the craters of Wabar, rather than being volcanic, were perhaps depressions caused by the fall of meteorites. He did not at that time know how few meteorite craters had been discovered in the world and did not in the least anticipate the great interest that would be evoked by the finding of "new" ones.

The rabbit-sized iron eventually reached the British Museum along with a number of smaller fragments. The metal was shown to be meteoritic, and both the slaggy material and the "black pearls" were found to be silica glass. The association of these materials with various other features of the circular structures definitely established Wabar as the site, not of ancient ruins, but of another group of meteorite craters. The color of the "black pearls" was later attributed to their probably having been ejected as particles of molten silica by the exploding meteorite into an atmosphere of vaporized nickel and iron.[6] And the camel-sized lump of iron was perhaps not a myth after all, for in 1966 T. J. Abercrombie reported having found a 4,800-pound iron meteorite at Wabar.

L. J. Spencer of the British Museum summed things up in 1933. He remarked that through the work at the Arizona crater, criteria had been established for the recognition of meteorite craters, and he listed the total number of more or less certain, possible, and probable meteorite craters then known as nine. The list included two single craters where meteoritic material had been found, Arizona and Odessa; two groups of craters where meteoritic material had been found, Henbury and Wabar; two possible locations, Oesel and Tunguska, groups of craters with no meteoritic material; and two doubtful ones, in Africa and Persia, single craters without meteoritic material. Ninth on the list were the Campo del Cielo craters in Argentina. They had been known for many years. In fact, native iron had been found near them by a Spanish explorer in 1576, and it was in this general region that Rubin de Celis discovered a mass of meteoritic iron in 1783. Nevertheless, an investigation of them by J. J. Nágera in 1926 resulted in the conclusion that they were man-made. In Spencer's opinion, however, there seemed to be no doubt that they were meteorite craters. Their meteoritic origin was confirmed in 1965,[7] and in 1970 a meteorite weighing eighteen tons and buried at a depth of about five meters was discovered in one of the craters.[8] By the early 1930s, interest in the subject was increasing to the point that meteorite craters were actually being looked for, and now and then a "new" one was discovered.

## CHAPTER 5

# *Explosive Impact*

NO DOUBT G. K. GILBERT'S interpretation of Meteor Crater would have been different if he had known that when a large meteorite traveling at enormous speed strikes the earth, the impact must cause vaporization of both meteoritic and terrestrial material. In part, his oversight came about because the interrelationship of the sciences was much less obvious a century ago than it is today. At that time, a wide separation between the various scientific disciplines was felt to be the normal state of affairs. Understanding of kinetic energy had been achieved several decades earlier, but it was the result of work by engineers, chemists, and physicists rather than geologists. Assimilation of its implications by scientists in general was by no means complete at the end of the nineteenth century.

It has been said that to many early investigators the mechanics of impact was a subject of despair. Understanding of it was obstructed in the early nineteenth century by the caloric theory of heat (although, earlier, heat had been equated with molecular motion). According to the caloric theory, heat is a weightless substance which flows from a region of high temperature to one of lower temperature much as water flows from a higher to a lower altitude. Doubts about this theory began to appear after Count Rumford's experimental development of heat by friction in 1798. In 1824 a young French engineer, N. L. S. Carnot, found during his work on steam engines that a complete transference of heat did not take place; some heat was transformed into mechanical energy. Carnot concluded that heat is the motion of particles of which materials are composed, but his work attracted little attention at the time.[1] During the next decades, however, investigation of the behavior of gases established the idea of heat as a form of motion. During the 1840s and 1850s, many investigators, particularly J. E. Mayer, H. Helmholtz, and J. P. Joule, showed that heat and "work" are equivalent and interconvertible. This idea developed into the principle of the conservation of energy, which has been called one of the great achievements of the human mind. In mathematical form it is familiar to students of physics as $e = \frac{1}{2}mv^2$; that is, the kinetic energy is equal to one half of the product of mass and the square of velocity.

Assimilation of this new and difficult concept by scientists in general was slow, but to some investigators its implications were immediately clear. J. P. Joule, for instance, remarked in 1848 that the ignition of meteors by violent collision with our atmosphere was an illustration of this principle, and in 1853 J. J. Waterston suggested that the heat of the sun might be due to the impact of meteorites falling upon its surface. Nevertheless it was many years before the principle was generally applied to the study of the mechanics of impact.

Gilbert remarked that when a meteorite strikes the moon, some of the lunar surface must be melted by heat produced by the impact. He overlooked the likelihood of vaporization of meteoritic and lunar material, although this was suggested by earlier writers. G. P. Merrill thought that some vaporization of the Canyon Diablo meteorite probably occurred during the impact that formed Meteor Crater, but he found no indication of intense heat other than fused quartz. In addition, he found it hard to accept the idea that heat as great as was indicated by his calculations could have been possible. But communication among investigators was slow, and opinions continued to differ. For example, Moulton's little-publicized reports suggested that heat sufficient to vaporize the meteorite existed during the formation of Meteor Crater. However, in 1930 A. F. Rogers, as we have seen, noted that the fused quartz indicated heat greater than that produced by volcanism and attributed it to friction. Although general acceptance of the fact that the impact of a high-speed meteorite upon the earth creates an explosion was slow, by the end of the 1930s it was widely accepted. And curiously enough, interest in it was aroused in the course of attempts to explain the scarcity of elliptical craters on the moon—the problem that Gilbert attempted to solve with his theory of moonlets.

In a discussion of the formation of meteorite craters at a meeting of the British Astronomical Society in 1915, A. W. Bickerton "pointed out that the normal meteoric speed is sufficient to produce an explosive action, and that consequently oblique impact will produce roughly circular rings."[2] 1916 E. Öpik demonstrated that, as kinetic energy depends upon the square of the velocity, the energy of a large meteorite traveling at a speed of many miles per second is enormous, and the energy released on impact must cause an instantaneous explosion which creates a round crater no matter what the angle of impact may have been. But Öpik wrote during the First World War, and in Russian, both of which circumstances were unfavorable to wide dissemination of his ideas.

H. E. Ives remarked in 1919 that one of the chief objections to the impact origin of lunar craters was their uniformly round shape. He noted that, as a scientific by-product of the First World War, the explosion of experimental bombs at Langley Field, Virginia, had resulted in craters "whose resemblance to the pitted surface of the moon will strike the most casual observer." He remarked that although it might seem far-fetched to liken meteors to explosive bombs, consideration of the probable mass of the meteor and its velocity on striking the moon indicated that tremendous heat must be produced by such an impact. "Even if we assume that nine-tenths of the heat is given up to the surroundings we still have . . . a temperature amply sufficient to gasify any known material, that is, *to produce an explosion.*" And this, he concluded, answered the perplexing question posed by the almost uniformly circular shape of the lunar craters, "for it is clear that the shape of the cavity has no reference to the angle at which the bomb strikes, but takes its form from the symmetrical explosive forces. . . ."[3] Nevertheless, many astronomers freely admitted their ignorance concerning the origin of lunar craters and the circular mountains which surround them, preferring to reserve judgment rather than adhere to either the volcanic or the meteoritic impact theory.

Independently formed conclusions were reached by another investigator, A. C. Gifford, in 1924. Gifford pointed out that G. K. Gilbert had in 1892 stated his belief that, since an iron or stony body falling upon the moon from an infinite distance and influenced only by the moon's attractive power would reach the lunar surface with a velocity of one and a half miles per second, the heat developed during such an impact would be considerably greater than that necessary to fuse the body. Gilbert had also noted that the average velocity of shooting stars was estimated at about forty-five miles per second, thirty times that of a body falling freely to the moon. He had come to the conclusion that when iron or stony masses traveling at such enormous speeds were suddenly arrested by encountering the moon, the energy released would cause the melting of the masses themselves and also of considerable lunar material. This melting of material, he had suggested, might explain the level surface of the inner plains of the lunar craters.

Gifford, too, noted that in the lunar environment, with little or no atmosphere to impede it, a meteorite or shooting star would strike the surface of the moon with its cosmic velocity undiminished. He pointed out that "the fact which has not been taken into account hitherto in considering the meteoric hypothesis is that a meteor, on striking the surface of the moon, is converted in a very small fraction of a second into an explosive compared with which dynamite and T.N.T. are mild and harmless."[4] In order to compare the energy of a meteorite with that of the best-known high explosives, Gifford gave the table below, which "enables comparison to be made as soon as the velocity of impact is known."[5]

Gifford suggested that in an impact of such violence, the meteorite would penetrate the lunar surface, where its tremendous energy, released in a confined space, must cause not merely the fusion of itself and the materials it encountered, but their instantaneous vaporization—and must inevitably result in a huge explosion. He gave a vivid description of his conception of such an event:

The explosion commences before the onward motion of the meteorite is stopped, and so the initial cavity widens out as it

---

### A. C. Gifford's Table of Energy in Calories per Gram

| Explosive | Energy in calories per gram |
|---|---|
| Tri-nitro-toluene (T.N.T.) | 924 |
| Tri-nitro-benzene | 940 |
| Tri-nitro-phenol (picric acid) | 914 |
| Tetryl | 1,094 |
| Nitroglycerine | 1,478 |
| Dynamite | (about) 1,100 |

A meteorite moving with a velocity of—

| Miles per second | Energy in calories per gram | Miles per second | Energy in calories per gram |
|---|---|---|---|
| 1 | 310 | 20 | 123,900 |
| 2 | 1,239 | 25 | 193,800 |
| 3 | 2,779 | 30 | 277,900 |
| 4 | 4,947* | 35 | 379,800 |
| 5 | 7,745 | 40 | 494,700 |
| 6 | 11,120 | 50 | 774,500 |
| 7 | 15,180 | 100 | 3,098,000 |
| 8 | 19,830 | 200 | 12,390,000 |
| 9 | 25,090 | 300 | 27,790,000 |
| 10 | 30,980 | 400 | 49,470,000 |
| 15 | 69,750 | | |

From A. C. Gifford, 1924

* In the original table the figure for 4 miles per second was 4,747, evidently a misprint.

deepens. The subjacent materials are forcibly compressed, whilst those above and around are shattered and pulverized and hurled aloft. In a few seconds a vast structure is fashioned which may preserve for untold geological ages the record of its birth.[6]

He pointed out that the angle of impact would not be important, for "whenever a meteorite penetrates into the crust, the marks of ingress are completely blown away by the resulting explosion. Whatever the angle at which the meteorite strikes the surface, the explosion acts radially outward."[7] According to Gifford's calculations, the theoretical form of the "vast structures" so created would be that of lunar craters.

Gifford also remarked that while a meteorite might strike the moon at cosmic velocity, the speed of a meteorite which reached the earth must be greatly diminished by its passage through the air, so that "it strikes the earth with so slight a momentum that it can penetrate only one or two yards into the ground, and may even fail to bury itself."[8] And indeed, this view seemed to be confirmed by many meteorites which had been found, or seen to fall, on the earth. In any case, Gifford did not apply his ideas of explosive impact to the problems of the few terrestrial meteorite craters then known, though he remarked in 1930 that "the best known example of meteoric action on earth is the Meteor Crater in Arizona which bears a strong resemblance to one of the smaller lunar pits."[9]

Although interest in meteorite craters was increasing, and although Gifford's paper was published in 1924, A. R. Alderman was apparently unaware of it at the time of his investigations in central Australia in 1931. He confidently expected the discovery of meteoritic metal beneath the Henbury craters. He employed a hand-operated boring tool in an attempt to locate meteoritic iron in one of the smaller craters. Traverses were made of the main crater to see whether the compass needle showed any deviation which might indicate the presence of a buried metallic mass, and the inconclusive results were attributed to shortcomings of the prismatic compass which was used. Alderman remarked that "until the position of any iron masses buried beneath the craters has been located, the direction of the fall of the meteoritic bodies must remain a matter of conjecture."[10] He was, however, acquainted with problems connected with the undiscovered meteorite at the Arizona crater, for he wrote, "If, as at the great Meteor Crater at Canyon Diablo in Arizona, magnetic methods prove to be somewhat unsatisfactory, it is highly probable that good results would be obtained by the use of gravimetric, seismic, or electrical methods."[11]

I. A. Reinvaldt, as we have seen, attributed the bowl shape of the larger Estonian craters to explosive action, which he explained as due to the sudden creation of steam by heat generated at the moment of impact. In 1932 a similar suggestion was made by L. J. Spencer in an attempt to explain evidence of explosive action at the Arizona crater:

> . . . the fusion of the sandstone, giving silica-glass, could no doubt have been effected by heat generated by the impact of the mass. With the development of a temperature (1400°– 1800°C) sufficient to fuse quartz, and the presence of water in the surrounding rocks, explosive action from this secondary cause must have helped in making the crater and scattering the fragmentary material. Such a "back-fire" would explain why the meteorite fragments are found only outside the crater, and it may also be the cause of the upward tilting of the strata in the crater walls.[12]

The following year, Spencer noted that the group of large irons excavated by R. Bedford from within crater No. 13 at the Henbury site was of special interest. It represented, he remarked, the first record of any large mass of meteoric iron having been found *inside* a crater.[13] Henbury crater No. 13 was reported to be about ten yards in diameter and about three feet deep.

Even L. J. Spencer, however, who was at that time perhaps the most widely informed authority on the subject of meteorite craters, remarked in 1933 that nothing was yet known of the mechanics of their formation. He added that they seemed not to be merely dents or holes made just by the projectile force of the meteorite, as previously supposed, but appeared rather to be explosion craters due to the sudden vaporization of meteoritic and terrestrial material. In fact, he calculated that (as $e = \frac{1}{2}mv^2$) 100 tons of iron traveling with a velocity of twenty-four miles per second would, on colliding with the earth, develop a temperature of 13,000,000°C. Even with a velocity of only ten miles per second, a temperature of about 200,000°C would be created. But Spencer experienced difficulties similar to those of G. P. Merrill twenty-five years earlier. He simply could not believe such figures. "This result seems absurd," he wrote, "and no doubt many factors have been overlooked in this simple calculation."[14]

At their first appearance, concepts completely outside the range of familiar experience are likely to be startling and hard to fit into preconceived notions. Today, ideas of enormous, unimaginable temperatures have become almost commonplace and are referred to so casually that it is not easy to reconstruct the feelings of astonishment and incredulity which, for most people in the early decades of this century, stood in the way of their acceptance.

L. J. Spencer was a geologist, an outstanding mineralogist, and equally notable as a petrologist. His work included descriptions of previously unknown minerals and of investigations over a wide field

of descriptive mineralogy. Through his system of labeling and cata-
loging, the mineral collection at the British Museum became known
as the best documented and indexed in the world. For fifty-five years
he edited the *Mineralogical Magazine,* and in this capacity he was
known for his meticulousness, his limitless industry, and his great
breadth of knowledge. Besides collections of minerals, he had in his
possession a collection of dictionaries in forty-one languages.

In 1927 Spencer succeeded G. T. Prior as Keeper of Minerals at
the British Museum, where he continued to work on the meteorite
collection that Prior had begun. His interest was further stimulated
by a visit to Southwest Africa in 1929. In a public garden at Windhoek
he was shown a pile of some thirty masses of meteoritic iron, collected
in the neighborhood of Gibeon in Great Namaqualand; and near
Grootfontein he examined the largest known meteorite, some sixty
metric tons of iron, lying where it fell on the Hoba West farm. In 1934
he took part in an expedition to the Libyan desert, where large chunks
of silica glass (lechatelierite) had been discovered by A. P. Clayton in
1932. Although Spencer suspected that the glass was formed by
meteoritic impact, no meteorite craters were discovered. The source
of the Libyan glass remains a mystery to this day.

Spencer's remark in 1933 that nothing was then known of the
mechanics of the formation of meteorite craters indicates that he was
unacquainted with Moulton's work, and he does not seem to have
found support for his "absurd" figures in the work of other investigators. It was not at all reassuring to him to find that other scientists tended to have opinions quite different from his—and from each
other's. As he pointed out, S. Mohorovičić (a distinguished seismologist, son of the Yugoslav seismologist whose name was given to the
discontinuity between the crust and mantle of the earth) claimed that
the explosions which created the lunar craters were volcanic. And
Alfred Wegener, whose concept of mobile continents profoundly influenced later geological thinking, performed experiments reminis-
cent of Gilbert's "knitting": he threw spoonfuls of cement powder on a
smooth surface of the same material and concluded that lunar craters
were the result of the percussive impact of meteorites.

By 1937 Spencer had developed a clearer picture of the events
which take place at the moment when a large meteorite, traveling at
high velocity, strikes the earth or the moon. To a great extent, this was
a direct result of his own mineralogical investigation of samples of
silica glass sent to the British Museum from Wabar and Henbury. The
silica glass must, he knew, have been formed by the melting of quartz
sand, for which a temperature of at least 1700°C was necessary. It was
plentiful at Wabar, in slaggy masses and "bombs"; and some of the
Wabar samples, beneath a black glassy exterior, consisted of frothy or
bubbly snow-white silica glass, similar in appearance to pumice.
Indeed, Spencer had already explained the bombs and "black pearls"

found at Wabar as formed when bits of molten silica were ejected into an atmosphere of vaporized silica, iron, and nickel.[15] Very little moisture was present in the desert sand at Wabar; therefore the highly vesicular structure of the silica glass indicated boiling and vaporization of the silica itself.

Small black spots within the black silica glass showed a metallic luster by reflected light; but as Spencer remarked, "the idea that they could be metal appeared at first to be so improbable that they were thought to represent small bubbles filled with the grinding powder used in the preparation of the section."[16] However, under careful microscopic examination, the black silica glass was found to contain minute spheres of nickel-iron, in some portions as many as two million per cubic centimeter. Spencer interpreted this finding as indicating vaporization of the meteoritic iron and nickel, and he pointed out that the boiling points of iron, 3200°C, and of nickel, 3377°C, are "calculated for the pressure of one atmosphere, but under the enormous pressures produced by the explosions at the meteor craters they must have been considerably higher."[17] The vaporized metals had condensed in a fine "drizzle," eventually forming minute metallic spheres within the silica glass. Their surfaces appeared brightly polished, indicating that they had not been oxidized and suggesting that the earth's atmosphere had been "blown away in a fiery blast."[18] Spencer concluded that all this must have occurred when the kinetic energy of the meteorite was suddenly transformed into heat. Such an explosion would eject all fragments of the meteorite from the crater, which would be circular whatever the angle of impact might have been.[19]

In addition to the evidence of violent explosions, Spencer had by this time found some support for his "absurd" figures. "The idea of such a gaseous explosion," he wrote, "was anticipated in a remarkable paper which has only now come to my notice, and which appears to have been generally overlooked."[20] The remarkable paper was that of Gifford, containing his table of energy in calories per gram for meteorites moving at various velocities.

On a less practical plane, improved understanding of impact mechanics offered support for the argument that lunar craters are the result of meteoritic impact and an explanation for the scarcity of elliptical craters on the lunar surface. But the explanation seemed to raise more questions than it answered. If lunar craters are caused by meteoritic impact, and if the earth and the moon have existed in their present relationship over vast periods of geologic time, why are there so many more craters on the moon than on the earth? Why are the moon's craters so much larger than those discovered on the earth? And why are there no terrestrial craters with central peaks?

These were perplexing questions. Perhaps, it was suggested, more meteorite craters exist on the earth than had been previously supposed. Perhaps the problem was just a matter of discovering them. Since the larger craters on the moon must be very old, as most of them appear to have suffered bombardment by smaller bodies, it was suggested that collisions of large meteorites with the earth might also have occurred more frequently in the remote past. On the earth, evidence of such events would in most cases have been obliterated by erosion; but some indications, if they could be recognized, might have survived. Geologists began to search for them. And before long, attention turned to peculiar, little-understood "cryptovolcanic" structures.

## CHAPTER 6

# Cryptovolcanic Structures

WALTER HERMAN BUCHER WAS a well-known American geologist whose professional career spanned half a century. His early training was in paleontology, but later he became intensely interested in the origin of mountain chains and other deformations of the earth's crust. For many years he stoutly maintained that various holes and depressions which were more and more frequently being called meteorite craters by other geologists had instead been created by processes originating within the earth. Curiously enough, in spite of his own vigorous support of a quite different interpretation, some of his work contributed substantially to recognition of the effects of meteoritic impact.

Bucher was born in Akron, Ohio, in 1888, but much of his early life was spent in Europe. After receiving his Ph.D. in geology from Heidelberg, he took a brief course in English and then returned to the United States. He attended lectures in geology and paleontology at the University of Cincinnati in order to acquaint himself with his native land and to improve his English, and eventually he became a member of the geology department. Many years later he recounted an incident which occurred during his early teaching experience in Cincinnati—an incident that profoundly affected his later work.

In the course of a lecture to a beginning geology class Bucher described local geological features, noting that the land surface in Ohio indicated intensive glaciation in the recent geologic past. He pointed out that much of the state is covered with sheets of glacial drift, composed of various layers of clay, sand, gravel, and pebbles. Great blocks and boulders of granite, diorite, quartzite, slate, and many other rocks foreign to the region are scattered about, lying

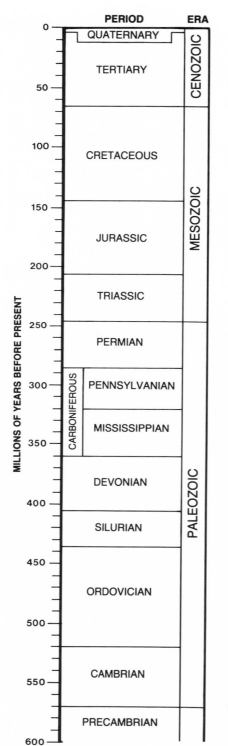

6.1   *Geologic time scale.*

where the melting glaciers left them. The glacial deposits rest directly upon ancient and often fossil-bearing sedimentary rock, formed in Paleozoic seas that covered the region more than three hundred million years ago; and beneath the glacial drift, flat-lying layers of Paleozoic rock extend unbroken as far east as the Appalachian uplift. In fact, as Bucher noted, in the great interior platform of North America the rocks are flat. Only in the marginal mountains—the Appalachians to the east and the Cordillerans to the west—are the rocks upraised and deformed.

Bucher was an excellent lecturer and a favorite with the students, and he was somewhat astonished when a young man who had apparently been listening with care asserted bluntly, "Professor, you are wrong!"

"Wrong?" repeated Bucher. "What do you mean?"

"Some of the rocks are not flat," the young man declared. "I know a place where the rocks are standing up on end. Less than a hundred miles from here."

"Perhaps it's a glacial formation," suggested Bucher. "Rocks might have been piled up by the ice."

The young man shook his head. "I live on a farm," he said. "I know what glacial drift looks like. This is different. I could show you!"

Bucher was very willing to be shown, and the eventual outcome of the incident was an excursion to a place near Sinking Spring, about fifty miles east of Cincinnati. Close to a corner where Adams, Highland, and Pike counties meet, the young man pointed out a hill in which the flat-lying strata of the surrounding countryside were indeed tilted and even up-ended. The hill was known as Serpent Mound, and Bucher became extremely interested in it.[1]

Serpent Mound proved to be an almost circular hill with a diameter of about four miles.[2] In the central part of it were great randomly tilted blocks of shale and limestone, in which Bucher recognized Ordovician fossils, indicating an age of more than 450 million years; and strangely enough, these rocks were some 350 feet above their normal regional position below the surface of the ground. In the central part of the hill was a small area, only a few hundred feet across, where he found slabs of rock bearing typical fossils of an even earlier stage of Ordovician time. This rock normally lay some 950 feet below the surface of the ground. In Bucher's opinion the cause of such astonishing vertical displacement in so small a space could only have been the driving force of an explosion. He observed that Serpent Mound was enclosed by a depression edged by a circle of hills; or, as he described it, "marginal anticlines lie outside the ring depression."[3] He later made the interesting remark, the significance of which was not apparent at the time, that "seen from a viewpoint on the edge of the escarpment which forms their edge, this circle of hills almost suggests a lunar crater."[4]

During his examination of Serpent Mound, Bucher was forcefully reminded of descriptions he had read of the Steinheim Basin in southern Germany. This was a circular depression about a mile and a half in diameter and more than 250 feet deep, and as E. Werner pointed out in 1904, it was puzzlingly different from any known geologic feature. Werner suggested that it might have been created by some great catastrophe such as a collision of the earth with a cosmic body—in other words, that it might be a meteorite crater; and in 1905 it was investigated by W. Branca and E. Fraas. They found that the intensely disturbed and fragmented rocks which formed a hill in its center were much older than those exposed in the rest of the basin, and that they had somehow been uplifted 500 feet or so above their normal level. The rocks in the hill showed evidence of some sort of violence, for they were jumbled, faulted, and fractured, and contained curious, distinctively striated rock fragments which they called shatter cones. Although they found no volcanic materials, Branca and Fraas interpreted the structure as having been formed by what they described as a kind of muffled volcanism, for which they coined the term "cryptovolcanic."

Not far from the Steinheim Basin was another puzzling and much larger depression, the Ries Basin, or Rieskessel (Giant Kettle), with a diameter of fourteen miles. Huge broken rocks lay in great confusion around its rim, and enormous boulders, apparently ejected

*6.2   Shatter cone from the Steinheim Basin, West Germany.*

by some tremendous explosive force, were found more than thirty-five miles from its center. Many suggestions were offered as to its origin, and opinions varied widely. A plentiful glassy material, now known as suevite, was interpreted as indicating some sort of volcanism, and the most widely accepted explanation was that the Ries Basin was the result of volcanic processes similar to those involved in the creation of maars.

Bucher mapped Serpent Mound in 1920. In 1921 he published a brief announcement pointing out that its structure was "strictly analogous" to that of the Steinheim Basin. The discovery of "new" evidence of cryptovolcanic action was of great interest to geologists, for until that time the Steinheim Basin was the only known example (with the possible exception of the Ries Basin) of cryptovolcanic structure. In 1924 Bucher made a trip to Europe and visited the Steinheim Basin. Its striking likeness to Serpent Mound was overwhelmingly evident. No doubt remained in his mind as to the similarity of their origins.

On his return to the United States, Bucher commenced investigation of another peculiar dome-shaped hill of apparently the same structural type. Under the auspices of the Kentucky Geological Survey he mapped Jeptha Knob, an isolated hill in Shelby County, Kentucky, halfway between Louisville and Lexington. It was some thirty miles from a belt of hills and projecting spurs known as "knob country," close to the gradually retreating edge of a high bluff. In its isolation, as Bucher remarked, it suggested an old volcano rather than an outlying remnant of the escarpment. His investigations, however, showed it to be neither.

In 1925 Bucher published as detailed a description of the stratigraphy, structure, and topography of Jeptha Knob as he felt to be possible in the face of the problems it presented. He remarked that

> exposures of the size and continuity required to define the boundaries of stratigraphic units with the precision customary in modern stratigraphic work are lacking almost completely within the disturbed portion of the Jeptha Knob area . . . it soon became evident, therefore, that the attempt to distinguish on the map the units of the Ordovician section in their accepted definition had to be abandoned. Only divisions based on fossil forms that are abundant . . . could be successful.[5]

Through study of their fossil content, Bucher identified various rocks of Ordovician age which, in the central part of the hill, he found to be some 200 to 400 feet above their normal position in the surrounding region. These rocks were steeply tilted in various directions, and in some places were almost vertical. Structural relations between them were impossible to determine. The highly disturbed central part of the hill was almost entirely enclosed by steeply dipping faults, and the whole structure was surrounded by a ring depression in which, as Bucher noted, the two points of greatest depth lay opposite each other on the north and south sides of the hill. Marginal synclines and anticlines, also, were best developed near these two deep points. Thus, Bucher remarked, "the whole suggests a crude symmetry in the distribution of stresses during the formation of the structure."[6] He noted, also, that the surface of the central part of the hill was formed of horizontal Silurian rock which rested unconformably on the disturbed and uplifted Ordovician layers. Below it in some places were breccias composed of angular fragments of all the various rock layers involved in the disturbance, as well as boulders of early Silurian age. He deduced from this that the disturbance, whatever it was, must have taken place during a relatively short period of early Silurian time.

Because all parts of the structure including the encircling hill and ring depression were "so intimately connected and so visibly complementary,"[7] Bucher concluded that they must be the result of a single disturbance; but because of the absence of cracks radiating from the center, the forces involved were probably not, he thought, of an explosive nature. He suggested that the structure might be the result of the slow rise of a rocky column about a mile in diameter, separated by fractures from the underlying crystalline rock, and "driven upward by an unknown force of deep-seated volcanic origin."[8] He speculated that in the analogous Serpent Mound structure, also, it was perhaps the pressure of rising magma that had raised the blocks in the center. He pointed out, nevertheless, that in Adams County, Ohio, as well as in Shelby County, Kentucky, the existence of subsurface intrusive rock is "purely hypothetical, with not one single petrographical observation to support it," and he went on to remark that "the complete absence of volcanic materials at the Steinheim and their practical insignificance in the much more remarkable 'Ries' basin were as much a puzzle to the German geologists as the similar American structure is to us."[9] In 1902 a German investigator claimed to have proved the presence of intrusive rock beneath the region in which the Steinheim and Ries basins lay, and Bucher hoped that a similar investigation might be possible "in the vicinity of at least one of our two cryptovolcanic structures."[10] But Bucher's work was attracting some attention, and the number of known cryptovolcanic structures in the United States soon exceeded two.

Descriptions of a peculiar feature called Upheaval Dome, in southeast Utah, caught Bucher's interest. It lay in a region where from time to time since the early 1890s wells had been drilled in search of oil.[11] Early results of drilling had been disappointing, but in 1920, oil in a test well south of Moab gave renewed impetus for exploration. At considerable depth this well passed through some 1,500 feet of salt. In the same year, showings of oil and gas occurred in another test well; and here, also, salt was encountered at depth. In 1925 a well on the Cane Creek anticline, near Moab, "blew in"; and oil and gas sprayed to a height of 300 feet before being brought under control. This well, too, passed through thick layers of salt. Drilling of various other test wells commenced, and in most of them salt was found at depth.[12]

A salt dome, or diapir, may be formed when a large deposit of salt, a relatively light material, lies beneath thick accumulations of heavier sediments. Over immense periods of geologic time, both salt and the surrounding sediments may slowly flow, in the manner of highly viscous liquids. Hydrostatic pressure may gradually force a column of salt upward, eventually piercing the overlying rocks and forming a hill at the surface. The existence of such hills, or salt domes, in southern Utah and southwestern Colorado was first suggested by T. S. Harrison in 1926. The only salt domes previously known in the United States were in the Gulf Coast region of Louisiana and Texas. As such structures often serve as traps for oil and gas, recognition of them in another part of the country was of great interest to the petroleum industry.

At least three locations where salt domes had been cut through by the Colorado River were observed by Harrison, and he predicted that many other domes of peculiar distortion would be discovered within the salt region. As an example, he mentioned the Christmas Canyon Dome, which exhibited a very unusual development. It was located in what seemed to be an old meander of the Green River, a tributary of the Colorado. He suggested that the removal of overlying rocks by stream action may have relieved pressure on salt deposits to such an extent that the salt gradually moved up along a path of least resistance, which in this case happened to be a highly developed system of jointing or fractures, indicated by the presence of rocks from a considerable depth in the "sharp and highly distorted crest."[13]

Harrison's paper included a sketch drawn by B. H. Parker in 1926, which showed a generalized cross section of Christmas Canyon Dome. This, along with evidence presented in a geologic map of the structure by E. T. McKnight, convinced Bucher that rather than being a peculiarly deformed salt dome, the structure was probably cryptovolcanic. McKnight, incidentally, referred to it as Upheaval Dome, the name it still carries. In support of his opinion Bucher pointed out that its outline was more or less circular; that the central uplift contained rocks raised some 1,200 feet above their normal level; and that it was surrounded by a ring-shaped depression. He noted, also, the presence nearby of an outcrop of igneous rock and a laccolithic mountain, which suggested that the processes responsible for the formation of Upheaval Dome might be connected with volcanic phenomena. In spite of Bucher's arguments, to many investigators the central uplift and ring depression presented an example of the theoretical form to be expected in a salt dome.[14]

In Bucher's 1933 list of cryptovolcanic structures then known in the United States, the largest example was the Wells Creek Basin, southwest of Cumberland City, Tennessee. In mapping the area for the Geological Survey of Tennessee, Bucher found it even more complicated than the other cryptovolcanic structures he had described. Ordovician limestones in its center had been raised at least 1,000 feet above their normal level and were broken into blocks which had been tilted and twisted into all sorts of positions. In many places the beds were vertical. The diameter of the central part of the structure was almost six miles. The complex marginal folds seemed extraordinarily wide, and the whole disturbance was found to be about eight and a half miles across. Noting indications of violent action, Bucher remarked that "aside from normal faulting, the pattern resembles that of damped waves, a central uplift surrounded by two pairs of down-and-up folds, with diminishing amplitude. It is the sort of pattern that results from a sudden impulse such as that of an explosion."[15] In addition, an explosive event was indicated both by great quantities of breccia, made up of angular fragments of dolomite and limestone ranging from blocks more than two feet in diameter down to the "granular matrix" that filled the spaces between them, and by the presence of numerous shatter cones, similar to those discovered at the Steinheim Basin by Branca and Fraas. Bucher explained them as due to mechanical shattering, possibly caused by the action of gases under high pressure.

Bucher observed two places where fragments of lower formations had been caught along a fault plane in rocks of higher levels. He remarked that

> the fragments of lower strata were apparently forced up from below while the fault gaped open for an instant. When these are considered together with the fragmentation and jumbling of blocks in the center, the local intense brecciation, and the curious shatter cones, the assumption of an explosion as the cause seems unavoidable.[16]

In an area near Decaturville, Missouri, which Bucher visited in 1932, Cambrian rocks more than 500 million years old had been brought to the surface by intense local deformation. Interpretation of the outcrop was extremely difficult, as most of the rocks involved proved to be unfossiliferous dolomites. He observed "intense internal brecciation of many dolomitic layers within the wildly confused beds of the central area."[17] An outcrop of highly quartzose pegmatite (a coarse-grained igneous rock) in the center of the structure was considered by some observers to be merely the top of a buried hill of Precambrian rock. This led to considerable controversy. Bucher, however, was convinced that the structure was cryptovolcanic.

A low hill in Newton County, Indiana, formed by a complicated rocky outcrop, was called the Kentland disturbance and was Bucher's sixth example of cryptovolcanic structure. For many years geologists had been puzzled by this outcrop. As early as 1883, Newton County, Indiana, was described by John Collett, state geologist for many years, as almost entirely covered with glacial drift "impenetrable to the geologist's eye"; yet near Kentland, at three locations known as the Means, McCray, and McKee quarries, rocky beds close to the surface extended over an area of more than 100 acres. Within the quarries some of the rock layers were very steeply tilted, and they contained peculiar nodules of "cone-in-cone" (nested cones of non-organic origin) which were interpreted as caused by pressure "while the rocks were in a plastic condition." Collett listed seven fossil forms found by his assistant, G. K. Greene, in rocks of the McKee quarry. He remarked that ". . . these fossils indicate that the rocks are Silurian and probably of Lower Silurian [i.e., Ordovician] age. The mass is too large to admit of explanation by its transportation during the ice period. . . . This quarry is a mystery. . . ."[18]

Interpretation of the disturbed rocks at Kentland was not the only problem facing early geologists in Indiana. In other parts of the state, too, outcrops of disturbed rock had been observed as far back as the 1860s. In many valleys and stream cuts, and in the valley of the Wabash River in particular, there were great cone-shaped masses of rock, thirty to forty feet high, from the tops of which the rock layers sloped outward and downward in all directions—a structural form known as quaquaversal. Frequently they were only partially exposed, leaving the puzzling question of their form beneath the all-concealing drift to the imagination of the geologist.

It was thought that the various uplifts had probably been formed at about the same time, as they contained similar Silurian fossils. In 1886 S. S. Gorby suggested that they were created during a great mid-Silurian upheaval which formed what he called the Wabash Arch; and he remarked: "That the position of the rocks at Chicago, Momence, Kentland, Delphi, Wabash, Huntington, and other points is due to the same cause, I have no doubt whatever."[19] He mentioned finding several species of lower Silurian (i.e., Ordovician) fossils at Kentland. These, he claimed, were invariably found between the rock layers and never embedded in the stone. The fossils were light and delicate forms which "would readily be carried upward through rents or crevices by the force of escaping steam or gas, such as would accompany an upheaval of the character indicated by evidence all along the Wabash Valley."[20]

The idea of such an upheaval was convincing to some geologists. For example, Maurice Thompson remarked in 1892 that "the 'arch' formed by this upheaval consists of a vast series of low bubbles or cones"[21] which he called "cones of upheaval," and which he considered to be related to the recently discovered Indiana gas fields. But there were other opinions. A. J. Phinney declared that information obtained from deep wells entirely refuted the possibility of a structural arch in northern Indiana, and in addition, that "the supposed axis of upheaval" was several miles from the most northerly point where gas was found. He suggested that cone-shaped masses of rock might be attributed to coral reefs on the Silurian sea-bottom. The hypothesis of the Wabash Arch was short-lived, but many other attempts were made to explain the uplifted and steeply dipping rocks of Indiana. They were variously attributed to cross-bedding,[22] to peculiar kinds of cleavage[23] (that is, the natural tendency of rocks to split), to some unusual volcanic process,[24] to much-broken folds in regional rocks,[25] and to "mud lumps" similar to those found at the mouth of the Mississippi.[26]

With some irritation, G. K. Greene, who had been Collett's assistant in 1883, noted the failure of geologists in general to recognize the Ordovician age of the outcrops he had observed at Kentland. He was also highly critical of the fact that the extremely steep dips at Kentland had been included by most investigators in the same general class as those found in other outcrops in the state. In 1906 he journeyed to Kentland and obtained specimens containing Ordovician fossils embedded in the rock in an attempt to settle the dispute arising from claims that these fossil forms lay only between the rock layers, and thus to establish the Ordovician age of the Kentland outcrops. In spite of these efforts, however, the Ordovician fossils found at Kentland did not receive wide attention. The problems presented by the "domes" and "dips" of the rocks of Indiana remained unsolved.

Time passed. By 1921 the presence of Ordovician fossils embedded in rocks at the Kentland quarry seems no longer to have been a matter of dispute. They were mentioned in a matter-of-fact way by A. F. Foerste, who described the occurrence at Kentland of a brownish-gray limestone containing several Ordovician fossil forms in addition to those originally identified by Greene, and he correlated it with the Black River (Ordovician) limestone of New York.

In 1922 a description of geological formations of Indiana was published by E. R. Cumings, who was evidently aware of Bucher's investigations at Serpent Mound. In his opinion, the presence of

Ordovician rocks in the Kentland dome indicated that the structure was the result of large-scale crustal movements or volcanism. "The possibility of cryptovolcanic action," he remarked, "such as described by Bucher (1920, unpublished) in an area near Cincinnati, Ohio, should not be overlooked in any further study of these domes."[27]

In 1927 and 1928 Cumings and R. R. Shrock published results of studies which, for the first time, revealed fundamental differences between the "dome" at Kentland and the quaquaversal domes, or cones, of the rest of the region. The high dips at Kentland were, they declared, undoubtedly due to some unexplained faulting or crumpling of the rock. But the cone-shaped masses of rock from which the strata sloped steeply downward were composed of limestone or dolomite. These structures, they declared, were fossil coral reefs. In fact, "embedded within the Silurian rocks of northern Indiana is a group of reefs, unrivalled in magnitude and geologic interest, yet until recently completely misunderstood."[28]

These investigations finally cleared away the controversies of the past and established the reef origin of the domes and dips of Indiana. The Kentland structure, however, remained unexplained. In 1929 R. R. Shrock and C. A. Malott identified fourteen Ordovician species, in addition to those listed by Greene and Foerste, in the rocks of the Kentland quarries. It was determined, also, that rock layers were exposed at Kentland which, less than three miles away, were to be found 1,500 feet below the surface of the ground. Groups of geologists and geology students visited the location, compensating somewhat, perhaps, for the previous lack of attention paid to what Shrock and Malott called "one of the most interesting and puzzling geological problems in Indiana."[29] Fossils were collected, and shatter cones (referred to by earlier investigators as cone-in-cone) were observed. Some of the sandstone was found to be intensely fragmented, and in places it had been pulverized. Under microscopic examination the resulting white powder was found to consist of small angular fragments of translucent or transparent quartz. All in all, the detailed

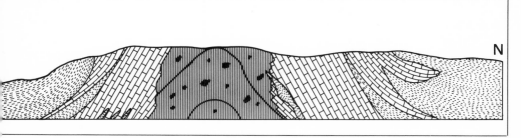

6.3    Cross section of the Wabash Dome, a typical reef structure in northern Indiana.

investigations convinced Shrock and Malott—and Bucher—that the disturbance was cryptovolcanic.

In 1936 Bucher published descriptions of the six "definitely identified American examples" of cryptovolcanic structure—Serpent Mound, Jeptha Knob, Upheaval Dome, Decaturville, Wells Creek, and Kentland—pointing out their similarities to each other and to the Steinheim Basin. These structures, he said, "must be considered unsuccessful attempts of volcanic materials to break through to the surface."[30] He suggested that when volcanic gases associated with an ascending column of lava were unable to escape, a deep-seated explosion eventually caused violent uplift of the overlying rocks. Forcible discharge of gas through cracks and fissures created during the process caused jumbling and shattering of the rocks. The result was a cryptovolcanic structure, consisting of a central uplift surrounded by a ring-shaped depression, which in some cases was edged by concentric hills, and in which there was no evidence of volcanism.

Cryptovolcanic structures were evidently not so rare as had at first been thought. Bucher noted that the cryptovolcanic nature of a structure at Flynn Creek, Tennessee, had been shown by C. W. Wilson and K. E. Born, and that a similar structure north of Howell, Tennessee, had been reported by the same investigators.[31] in 1933 P. B. King described an interesting structural feature near the Glass Mountains in Texas. This was the Sierra Madera dome, a circular group of hills about three miles in diameter, which rose to a height of about 600 feet above the surrounding plain. King noted that the strata in the hills appeared "to be jumbled and twisted in hopeless disorder. . . . Rocks within the compass of but a few miles have been uplifted thousands of feet and exhibit evidence of great compressional deformation, though to the encircling areas the beds are nearly flat."[32] King, perhaps unaware of Bucher's work, remarked that it invited comparison with the Vredefort dome, a puzzling structure in South Africa which was described in 1925 by two prominent South African geologists, A. L. Hall and G. A. F. Molengraff.

Although Bucher's explanation of cryptovolcanic processes and structures was widely accepted, there were misgivings on the part of a few geologists who felt that it did not adequately account for the tremendous explosive forces which must have been involved in the uplifting and shattering of huge masses of rock. Others did not believe that volcanic explosions were likely to occur without the production of volcanic materials, and a few voices were raised in disagreement. Among them was that of H. P. T. Rohleder.

As we have seen, the Ries and Steinheim basins of southern Germany were the first structures to be interpreted as cryptovolcanic;

that is, created by some sort of muffled volcanic explosion. Material thought to be of volcanic origin was discovered in the Ries Basin, but at Steinheim no evidence of volcanic activity was found. H. P. T. Rohleder, a noted geologist who lived for many years in Africa, refused to accept the idea of volcanic explosions which exhibited no trace of volcanic materials. He suggested in 1933 that the Steinheim Basin was a meteorite crater. He remarked that his attention had been drawn to a new publication by J. Kaljuvee, of Talinn, in which "this Estonian teacher, as clever as bold, tries to explain many geological features of the earth's crust through impacts of large meteoric bodies. As such he not only regards the Steinheim Basin, but also the Ries of Nördlingen, in spite of its vast size."[33] Rohleder described some of Kaljuvee's ideas as "extremely bold," but was nevertheless in complete agreement with his opinion as to the origin of the Steinheim Basin.

Rohleder noted similarities between the Steinheim Basin and a feature known as the Pretoria Salt Pan in the Transvaal, South Africa. Both were circular depressions enclosed by rims of coarse breccia and great blocks of broken rock, and no volcanic material had been found in either of them. It seemed more likely to him that they had been created by meteoritic impact than by any sort of volcanic process, although meteoritic materials were absent at both locations. As a possible explanation of this absence he suggested that the meteorite involved might have been of the stony variety, in which case all fragments of it would have disappeared long ago.

In 1936 a new and startling interpretation of cryptovolcanic structures was offered by J. D. Boon and C. C. Albritton, Jr. They pointed out that, as the rock layers in the rims of meteorite craters dip radially outward, the impact and explosion of a large meteorite is evidently capable of producing extensive deformation of the affected rocks—capable, in fact, of producing geologic structures. The dip of the strata away from the center might therefore indicate that the underlying rock layers had been domed by the event, due to elastic rebound after the enormous compression of the impact, and that

> as a result of impact and explosion, a series of concentric waves would go out in all directions, forming ring anticlines and synclines. These waves would be strongly damped by the overburden and by friction along joint, bedding, and fault planes. The central zone, completely damped by tension fractures produced by rebound, would become fixed as a structural dome. The general and simplest type of structure to be expected beneath a large meteorite crater would, therefore, be a central dome surrounded by a ring syncline and possibly other ring folds, the whole resembling a group of damped waves.[34]

Because rock layers to a considerable depth would be affected by the impact and explosion of a meteorite, the appearance presented by a meteorite crater at any given time would depend upon the extent to which the surface layers had been removed by erosion, as shown by Boon and Albritton's accompanying figure. They pointed out that

> it is only in the initial stage (along profile AA) that the crater clearly reflects its origin in the rim of ejected material, silica glass, and meteorite fragments distributed about it. The scar will become inconspicuous when the country is denuded to the level BB. When the area is down to the level CC the underlying structure begins to appear, and when the depth DD is reached the central uplift and ring folds become apparent. Should erosion proceed to depths below those affected by the meteoritic disturbance, the scar would be obliterated. On the other hand, if the scar should be submerged and covered with sediments, it might be preserved and subsequently revealed in the course of regional uplift and erosion.[35]

Cryptovolcanic structures were therefore strikingly similar to formations which might be expected beneath large meteorite craters after the surface features had eroded away. In addition, the bilateral sym-

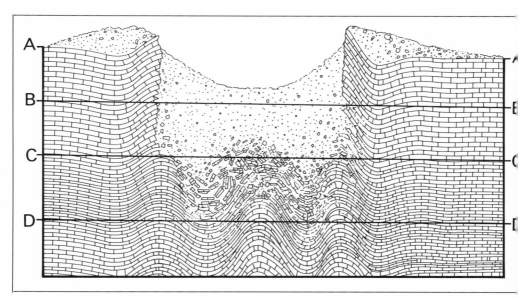

6.4  *Ideal cross section through a meteorite crater, as suggested by Boon and Albritton, Jr., in 1937. At AA, the crater is clearly visible. After erosion to BB, the scar is not readily apparent. At the levels of CC and DD, the central dome and ring folds become recognizable.*

metry observed by Bucher in several cryptovolcanic formations might be explained by oblique impact of a colliding meteorite, whereas an upward-directed explosion such as was postulated in cryptovolcanic processes would probably have produced a radial symmetry. As for the presence of meteoritic materials, considered by many people to be the final proof of meteoritic origin, "these most convincing clues," Boon and Albritton wrote, "are just the ones which will disappear first under the onslaught of weathering and erosion."[36] Nevertheless, in the absence of meteoritic materials most investigators found it difficult to argue that a particular structure had been formed by a meteorite. As Boon and Albritton remarked, one hardly knew which was better—a meteoritic hypothesis without meteorites or a volcanic hypothesis without volcanics.[37] Most scientists at that time favored the volcanic explanation.

Boon and Albritton suggested that three geologic structures which had tentatively been attributed to other origins had probably been created by meteoritic impact. One was the Flynn Creek structure in Tennessee, interpreted by C. W. Wilson and K. E. Born in 1936 as cryptovolcanic. It was an ancient explosion crater in which much breaking up and pulverization of the rocks had occurred, and where the rocks in the central part of the uplift had been raised some 500 feet above their normal level. The bilateral symmetry of the structure, and the fact that the rocky beds had been overturned on one side only, convinced Boon and Albritton that it was a meteorite scar. In their opinion, the somewhat similar bilateral symmetry of the Sierra Madera dome in Texas, as described by P. B. King in 1930, identified it too as a meteorite scar. King compared the Sierra Madera dome to the Vredefort dome in South Africa—an enormous circular uplift of subsurface rock some fifty miles in diameter. Boon and Albritton claimed that this great structure, surrounded by a girdle of tilted and overturned rocks, could be accounted for as the result of the impact and explosion of a gigantic meteorite. In support of their opinion, they noted that microscopic examination of rocks of the Vredefort district indicated that they had been subjected to enormous pressures.

The speculations of Boon and Albritton were published in a somewhat obscure journal. It was not until a decade later that they began to receive the attention of several investigators of meteorite craters.

## CHAPTER 7

# The Work of Harvey Harlow Nininger

EARLY CONTRIBUTIONS TO THE recognition of meteorite craters were made by investigators who were often widely separated and sometimes unaware of one another. The work of Harvey Harlow Nininger is now well known, but apart from assistance by his family his early investigation of meteorites and meteorite craters was carried out alone.

In 1923 Nininger noticed an article in the *Scientific Monthly* which began with the following paragraph:

"Meteorites, being solid bodies which come to earth from celestial space heated to dazzling brilliancy on the surface from friction with the air and detonating with terrifying intensity as they set up violent atmospheric compression and rarefaction waves, would appear to carry with them proofs of their extramundane origin as incontestable as could well be demanded. Yet it was not until the closing years of the eighteenth century that the scientific world became convinced of the reality of these objects and that they did actually "fall from the sky." It is said that even President Thomas Jefferson, progressive though he was in the science of his day, was skeptical about such occurrences, replying to the assurances of a friend that the actual fall was well attested by two Yale professors that he "would prefer to believe that two Yankee professors would lie rather than that stones could fall from heaven."[1]

The fall referred to occurred in Weston, Connecticut, in 1807, and a detailed account of it by Professors B. Silliman and J. L. Kingsley of Yale University appeared in 1809. The article in *Scientific Monthly* was written by A. M. Miller, of the University of Kentucky, who went on to describe some of the problems involved in locating a meteorite after an observed fall—"meteorite hunting," as he called it. He discussed a number of notable meteorites, including Ahnighito, the largest of the Cape York irons found in Greenland and brought to the American Museum of Natural History in New York by Admiral Peary in 1897; the Williamette iron (Miller's spelling; the more usual form is Willamette) found in Oregon in 1902 and brought to the same museum after a legal battle as to its ownership; the great Mexican irons on exhibit in Mexico City; and the very large one known as Bacubirito, which still lay where it was found in the state of Sinaloa, Mexico, and was excavated in 1892 under the direction of H. A. Ward.

No doubt Miller's article was read with interest by many people, for at that time little was known about the subject; but no one, perhaps, was more deeply stirred by it than Harvey Harlow Nininger. In fact, he remarked later that he could not remember ever reading anything that so completely captivated him.[2]

At McPherson College in Kansas, Nininger taught biology and conducted a field course in geology, but his interests were broad and encompassed the large domain known as natural history. It was with understandable chagrin, therefore, that after reading Miller's article he realized there existed a branch of natural history of which he was completely ignorant. He pondered over the subject of meteorites for many weeks and became increasingly curious about what a study of them might reveal. As samples of material from beyond our atmosphere they must, he felt, offer unique sources of information to the astronomer and the geologist. He set out to learn what was known about them and was soon dismayed to discover that even among astronomers and geologists ignorance as profound as his own was not uncommon. He spent considerable time calculating how, if one were fortunate enough to see the fall of a large meteorite, one might determine its course and the approximate location of its landing place, though there seemed little likelihood of ever putting such calculations to practical use. As he knew, the chances of witnessing such a spectacle were not large.

Remarkably enough, an opportunity soon presented itself. On November 9, 1923, scarcely three months after he first read Miller's article, Nininger and a friend were walking home after attending an evening program at the college. Suddenly, in Nininger's words, "a blazing stream of fire pierced the sky"[3]—and vanished. Nininger immediately bent down and plotted the direction of the fiery stream, as it had appeared to him, on the sidewalk, announcing as he did so that he was going to find that meteorite. This event reinforced the effect of Miller's article and changed Nininger's life.

Following a procedure suggested by Miller, Nininger lost no time in asking editors of principal newspapers in the state to publish a request for information from anyone who had seen the meteor. The request was quite precise. The desired information included the location from which the observation was made, the direction of the meteor (with respect to the observer) when it disappeared, and its height above the horizon (in degrees, vertical being 90°) at the moment of its disappearance. Replies were numerous and came from all parts of Kansas, Oklahoma, and Nebraska; but in most cases, unfortunately, precision was lacking. Evidently, to many people the meteor had appeared to be very close and to have disappeared—and no doubt subsequently to have landed—behind a nearby barn or just over the next hill. The mass of reports soon proved to be a welter of conflicting statements.

As Nininger remarked, the sight of a great fire-ball is an unexpected and exciting and possibly even terrifying event. The observations he received had been made not by scientists but by startled laymen and had been recorded a day or two after the event. Their great diversity was therefore not surprising. He patiently sifted through the mass of reports, and by plotting on a map the information given by what seemed to be the best half dozen of them, along with his own observations, he became convinced that the fall had taken place within a fairly constricted area perhaps forty miles across. He concluded that the meteorite, or a fragment of it, probably fell in the vicinity of Coldwater, Kansas, and another fragment possibly near Greensberg.

Compared with the great area over which the meteor was seen, a constricted band forty miles wide was very small; but as a location to be searched for a meteorite, which would probably look like a small rock, it was enormous. During the following year Nininger made several trips to the neighborhood of Coldwater, Kansas. He visited schools and newspaper offices, explaining the purpose of his search, as well as something of the nature and behavior of meteorites and the reasons for their value as objects for study. And he offered to pay a good price for any specimen found.

Although the meteorite of November 9, 1923, was not discovered, Nininger's efforts had some unexpected and interesting results. During one of his first visits to the area in which he suspected the meteorite to have landed, he was allowed to purchase a forty-one-pound meteorite which had been turned up by the plough some four years earlier, and which had since lain as a curiosity in a back yard. In the same neighborhood another meteorite was found a year or so after

Nininger's visit. It was at first thought to be part of the fall of November 9, 1923, but was later determined to be much older. This one, too, was turned up during ploughing.

Meteorites, as Nininger knew, were rare objects. Consequently, information about them was extremely limited. Reflecting on the matter, he wondered whether their scarcity might be related to an almost complete lack of information about them on the part of the general public. Two previously unknown meteorites had been found in the region of Coldwater where, as he said, he "had cause to believe still a third meteorite lay undiscovered"[4]; and they had come to light not because of the fall he had witnessed, but because he had alerted the public. Meteorite falls, after all, must have been occurring for many centuries. Perhaps in any locality a careful search might lead to the discovery of "new" finds. The only reason that meteorites were not found more frequently, Nininger decided, was that people did not recognize them. He determined to change that state of affairs.

Nininger embarked on a program of intensive searching for meteorites, at the same time disseminating information about them to anyone who would listen. He obtained the cooperation of newspaper editors. He gave lectures in schoolhouses and civic clubs, and informal talks in general stores, on the street, in bars, in fields, anywhere at all. He carried with him small meteorites for people to hold and examine, and he offered to buy any that might be found. He traveled many miles, sometimes to remote places, following tenuous and often disappointing leads.

People listened to his lectures. Occasionally someone would be struck by similarities between the meteorite samples he displayed and some stone they had noticed as being particularly heavy or of a curious shape. And sometimes such stones indeed proved to be meteorites. In time, meteorites began to appear from all sorts of odd places. One was brought to him which for twenty-five years had held down the lid of a pickle jar. Another had served as a doorstop. An iron meteorite had been used as a weight on the roof of a chicken house. Still another had plugged a rat hole in a basement. Nininger received them eagerly and paid for them on the spot. Four years after hearing him lecture, a boy who had then been in the sixth grade found a small meteorite in a cornfield. Nininger's prompt payment for this find was duly noted by the boy's neighbors. Among them was a young man who immediately produced a seventeen-pound meteorite. He had found it a year before, but because he feared his family's ridicule he had hidden it under a granary.

Nininger became more and more engrossed in searching for and studying meteorites. With the aid of his wife, who accompanied him on many expeditions, he gradually accumulated an impressive collection of them. Although the income derived from their sale, mainly to museums and other institutions, was extremely precarious, and although there were three Nininger children to support, he eventually gave up his position at McPherson College in order to devote all his time to meteorites.

Today, Nininger's great interest in stones and metal which fall from the sky is shared by many scientists, but in the 1920s and 1930s little attention was given to such matters. Nininger's approach was felt by many people to be unorthodox, to say the least, and grants of money to assist his work were almost impossible to obtain. Even G. P. Merrill, of the U.S. National Museum, who a couple of decades earlier had investigated the Arizona Meteor Crater in the company of Barringer and Tilghman, was unwilling to offer financial support for his program. He seems, in fact, to have had little interest in it. Nininger quoted Merrill as remarking that he thought meteorites had already given him "just about all they have to offer by way of information."[5]

Although little financial aid was obtained, local people became very interested in Nininger's work, and friendly assistance sometimes appeared in other forms. For instance, he was made curator of the Colorado (later Denver) Museum of Natural History, a post which entailed little in the way of duties beyond the exhibiting of his meteorite collection, but which carried a small honorarium. The owner of a fleet of trucks hired him as a driver with the understanding that whenever possible he might combine his own travel needs with those of the trucking company. Interest-free loans were often made available to him, and a local bank extended credit which made possible the purchase of expensive specimens from which sections were later cut, polished, and sold. Individuals who shared his enthusiasm occasionally provided financial backing for a particular expedition with the agreement that the resulting finds, if any, would be equally shared. As his work became known, opportunities to give lectures at universities and other institutions became increasingly frequent; and although the fees he received were small, they were a welcome addition to the tight family budget. All this was helpful and encouraging, though it hardly took the place of a dependable income; but somehow, in spite of financial problems which at times were almost overwhelming, Nininger managed to carry on his work. As F. L. Whipple has pointed out, the Nininger family was the first ever to survive by finding, collecting, trading, selling, and exhibiting meteorites.[6]

The treeless plains of west Kansas are covered with sandy loam which in many parts of the region is devoid of rocks; yet, in the early days of cattle raising, cowboys occasionally came across heavy black stones projecting from the soil. They were unusual enough to attract attention, and the cowhands are said to have used them for weight lifting. When agricultural operations began in the early 1880s, farmers and ranchmen sometimes received a severe jolt when a farm implement encountered one of these black stones—usually to the detriment of the implement. Such stones seem not to have aroused more than passing curiosity; but in an otherwise rockless region they were found to have practical uses such as holding down haystacks or the lids of rain barrels, as ballast for roofs of dugouts, or to anchor fencelines.

In 1885 a young couple named Frank and Eliza Kimberly[7] settled on a homestead in Kiowa County, Kansas, near Brenham. Eliza Kimberly immediately noticed a similarity between the heavy black stones which were now and then found on her land and a large meteorite which she remembered having seen as a child in Iowa when it was being taken to a museum. In spite of ridicule by her neighbors and indulgent teasing by her husband she gathered the heavy stones as they were found, and gradually accumulated a pile of them. She made many unsuccessful attempts to interest various scientists in her collection. Eventually, after interest had been stimulated by the fall of the Farmington meteorite in 1890, F. W. Cragin of Washburn College, Kansas, and F. H. Snow, of the University of Kansas, responded to her request to examine the stones, with the result that her suspicions as to their meteoritic origin were confirmed. About half of Eliza's rockpile was purchased on the spot, and the several hundred dollars she received in payment was (in those days) almost equal to the price of a farm.

After this advantageous transaction, the Kimberly homestead was increased by additional land, on which more "black stones" were discovered. The place became known as the "Kansas meteorite farm." It was soon visited by other scientists who also purchased specimens. The black stones, it was concluded, originated long ago in what became known as the Brenham meteorite fall (named as was customary after the nearest post office), when a large meteoritic mass broke up as it neared the earth, and meteoritic fragments were scattered over the region. Most of the specimens belonged to the class of meteorite known as pallasites—masses of nickel-iron with a sponge-like form, the cavities of which contained greenish crystals of olivine—similar to the one found long ago in Russia by Pallas. Other specimens were composed entirely of nickel-iron, and still others contained patches of both nickel-iron and pallasite structure. The "discovery" by scientists of the meteoritic rocks of Kiowa County, Kansas, resulted in the acquiring of specimens by several universities. It was suggested, incidentally, that ear ornaments made of meteoritic iron, and found some years earlier in prehistoric mounds of the Ohio Valley, had been fashioned from fragments of the Brenham fall.[8]

Interest in the black stones gradually waned. After selling about a ton and a half of the meteoritic rocks, the Kimberlys were unable to

7.1    *Excavation of a wall of the Haviland, Kansas, meteorite crater.*

find any more buyers, and when in the mid-1920s Nininger appeared at the "meteorite farm" there still remained specimens which he was able to acquire. A friendship sprang up between Nininger and the Kimberlys, and in the course of a later visit, in 1929, he chanced to hear that several meteorites had been found near an old buffalo wallow. He had noticed the wallow, a more or less circular depression, the first time he saw the farm, and the fact that meteorites had been found near it caught his immediate attention. During his scouring of the countryside in search of meteorites, the apparent absence of any formation which might be the remains of a meteorite crater had puzzled him, and he was constantly on the alert for indications of such a thing. The controversial Arizona and Odessa craters were the only formations he knew of for which meteoritic origins had been suggested.

The Kimberlys explained that the buffalo wallow tended to retain water longer than the surrounding land, and Eliza remarked that she had picked up about a bushel of small meteorites around it. Attempting to fill it in, they had ploughed and planted over it. They showed Nininger a wheat field within which a shallow elliptical depression was clearly discernible. He noted with great interest that the depression was edged by a low but quite distinct rim. In spite of some trampling of the wheat, Nininger was permitted to dig into the rim, where a few inches below the surface he immediately discovered a number of small, oxidized meteorite fragments.

In the 1920s, as we have seen, the idea that meteorite craters might exist on the earth was still regarded by some people as a fanciful product of unbridled imagination on the part of scientists. Even among geologists, the opinion that the Arizona crater had been formed by some sort of volcanism—a "steam explosion," as suggested by G. K. Gilbert—had not disappeared, although the possibility of the existence of terrestrial craters formed by meteoritic impact was receiving increasing attention. Nininger read Bibbins's description of the Odessa crater and later traveled to Texas to see it. His observations as to its shape, along with several small meteoritic fragments which he picked up during his examination of its rim, left him convinced that it was of meteoritic origin. Alderman's account of the Henbury craters appeared in 1932 and was soon followed by Philby's discovery of Wabar. These in turn were followed early in 1933 by L. J. Spencer's summing up of the then-current state of knowledge concerning me-

7.2   Clyde Fisher (left), Curator of Astronomy at the Hayden Planetarium in New York City, examines a meteorite with H. H. Nininger at the Haviland crater.

7.3    Clyde Fisher on a visit to the Haviland meteorite crater during the 1930s.

teorite craters. Nininger read of these developments with intense
interest and with an increasing desire to excavate the Kimberlys'
buffalo wallow. Finally, in 1933, he was able to complete arrangements to do so. And on this occasion some financial support for the
undertaking was obtained from the Colorado Museum of Natural
History, due no doubt to the fact that the director of the museum,
J. D. Figgins, who cooperated in the work, was keenly interested in
meteorites. He had been an assistant to Admiral Peary on the three
expeditions to Greenland in the late 1890s when the Cape York
meteorites were transported to New York.

Excavation of the buffalo wallow was undertaken with great care.
Nininger was assisted by interested friends and neighbors, and news
of the project spread as far as New York City resulting in a visit to the
location by Clyde Fisher of the Hayden Planetarium. The work was
fully justified, for the buffalo wallow proved indeed to be a meteorite
crater. Elliptical in outline at the ground surface, it measured about
thirty-five feet by fifty-five feet (10 m by 17 m). Careful digging
revealed that the distribution of the meteorites, most of which were in
the form of small oxidized fragments, outlined the shape of the crater;
that is, "the specimens were distributed so as to conform to the
surface of an oblique inverted cone, with their size increasing as the
work proceeded downward. . . ."[9] Oxidized meteorite fragments
recovered from the crater were so altered by weathering that they
could have been mistaken for iron concretions, and they weighed
altogether some 1,200 pounds. During later investigations completed in 1927, Nininger obtained a large quantity of nickel-iron
particles by screening and magnetic combing of soil from the crater
area. In his opinion the recognition and excavation of the Kimberlys'
buffalo wallow, which became known as the Haviland crater, was the
most important discovery having to do with meteorites that had been
made in Kansas.

The curious fact that up to that time more meteorites had apparently fallen in Kansas than anywhere else in the country had been noted by various investigators. Some even suggested that the Kansas region might possess some peculiar quality of attraction for meteorites. Nininger did not accept this view. In his opinion, one reason for what seemed to be an abnormally large number of meteorite finds in Kansas was the absence in many parts of the state of terrestrial rocks. Meteorites stood out much more noticeably than if they were lying in rocky terrain. Another was the somewhat arid climate, in which disintegration due to weathering took place much more slowly than would be the case in a humid atmosphere. The nature of the soil and the extent to which it was cultivated probably also had an influence, but in Nininger's opinion, the most important reason was what he called the "interest factor." He pointed out that after a spectacular event such as an observed fall, heightened public interest often resulted in additional finds; for, as he remarked in 1933, "our knowledge of such falls depends upon the presence of human beings intelligent enough to notice the difference between meteorites and ordinary stones and interested enough to report them."[10]

It has also been pointed out that the number of meteorites found in the part of the country where Nininger actively alerted the public was remarkably high. A heightened awareness of the existence of meteorites, to which this success may be attributed, was a direct result of his work and reflected his own consuming interest. Knowing as he did that meteorites often break up as they approach the earth, he frequently made door-to-door inquiries in some locality where a single meteorite had been found and often discovered additional fragments of the fall. Through his efforts, starting in 1924, Kansas began to be "meteorite minded," as he called it; but he pointed out that "it required hundreds of free lectures to many thousands of students, farmers, and farm wives to bring about the thirty-four actually verified finds in the state which resulted from our program during the years 1923–1948."[11] Although Nininger has been described as the man who collected more meteorites than any other man who ever lived, it dismayed him to be viewed as a mere collector. While collecting indeed occupied much of his time and contributed substantially to his livelihood, he did not regard it as his main purpose. As he remarked in 1972, "collecting served as a sort of platform or footing on which to stand while I sought to educate, and while I pleaded constantly for an organized program of meteoritical research."[12]

Much larger than Haviland, the Odessa crater, some ten miles southwest of Odessa, in Texas, was almost 600 feet (183 m) in diameter. Nininger visited it in 1932, supplied with what he described as a magnetic rake, with which he collected some 1,500 meteoritic fragments weighing altogether about eight pounds. He made another visit in 1935, armed with a magnetic balance, a device invented by a man who planned to search for some of the fabled buried treasure of the Southwest. This time Nininger's harvest consisted of a total of twenty-seven meteorites, with a combined weight of some thirty-four pounds.

Nininger thought that a mass of meteoritic metal, such as was believed by many people to lie below the Arizona crater, must also exist below the Odessa formation, too deeply buried to be revealed by the instruments he had used. He hoped that excavation would be undertaken by the federal government so that all possible means might be taken to preserve the surface appearances. Unfortunately, Nininger's recommendations had little effect. The random and careless digging which later marred the Odessa crater has been lamented by many scientists.

In 1936 Nininger gave a lecture at the Adler Planetarium in Chicago. In the audience was F. R. Moulton, who only a few years before had delivered his discouraging report on mining prospects at the Arizona crater. After the lecture he introduced himself to Nininger and expressed great interest in his work. When he discovered that it was carried on with no financial support, he suggested approaching various institutions on Nininger's behalf when he visited Denver later that year. This he did, but in spite of his great prestige and the fact that he had the support of many other scientists, the various ideas and plans which he proposed came to nothing.

In the course of a luncheon conversation during his visit to Denver, Moulton remarked that Nininger still believed that a meteorite lay buried in the Arizona crater. In reply, Nininger referred to Barringer's opinion, and to the report that a great mass, probably meteoritic, had been encountered during drilling.

"Nininger, there cannot be a meteorite in that crater!" Moulton declared.[13] He went on to explain that it was mathematically and physically impossible for a fast-moving mass large enough to have produced such a crater to be suddenly stopped and yet to remain intact. Instead, he insisted, it would explode.

In spite of his admiration for Moulton, Nininger—who described himself as a field and laboratory man rather than a theorist, and who did not at that time know of Moulton's 1929 reports—was not convinced. He had, after all, found many meteorites in the course of his work, some of them quite large. Although the fact that the Haviland crater yielded nothing but small fragments had caused some doubt to creep into his mind, it did not seem impossible to him that a mass of meteoritic metal might eventually be discovered beneath the Arizona crater. Nevertheless, he had great respect for Moulton and could not lightly dismiss his opinion.

It occurred to Nininger that evidence which could prove or disprove Moulton's theory might exist in the soil close to meteorite craters. He mulled over this idea for some time, and finally in 1939 he managed to obtain sufficient funds to undertake a search for small fragments around the Arizona crater. He designed a magnetic rake to be drawn behind his car, and with this instrument and some assistants he combed areas within two and a half miles from the crater which were free enough of boulders and scrub to be suitable for the use of such equipment—about twenty-three acres in all. Some 12,000 small nickel-iron fragments with a total weight of forty-two and a half pounds were collected, and careful records were kept of the locations in which they were found. A pattern of radial distribution seemed to indicate that they had been thrown from the crater.

Nininger's magnetic rake survey prompted further investigation. Among interested scientists there had been some discussion of the strange absence at the Arizona crater of plentiful silica glass such as that found at Henbury and Wabar. There was also speculation as to whether condensation droplets of nickel-iron such as those described by Spencer might exist there. Investigation of the soil in and around the crater seemed indicated, and this was a project for which Nininger was well prepared. During various explorations of the Arizona crater he had often carried a cane with a magnet attached to its tip, which was used to collect small metallic fragments from the ground. He was greatly interested in the dust and minute particles which clung to the magnet and had saved many samples of them in carefully labeled envelopes for future investigation. Quantities of similar tiny particles had also been gathered by the magnetic rake; and samples of this material, too, had been saved for examination when time might permit.

The war years imposed severe restrictions on travel. Meteorite hunting and meteorite research had to be laid aside. For a time Nininger worked for the War Production Board, scouring old mining towns for useable metal. Later he took a job with an oil company and engaged in exploration for new wells. When eventually he was able to resume his own work, the family decided to move to Arizona. The summer of 1946 was spent in Flagstaff, Arizona, while they looked for a suitable location for setting up a meteorite museum. During this period, Nininger and his wife and daughter made frequent trips to the crater. He became intensely interested in the small particles which collected on their magnets and spent many hours examining them. He began to suspect, with mounting excitement, that he was "finding the condensation products which Merrill had declared were lacking, and which the Moulton explosion theory seemed to require for its verification."[14]

In the fall of 1946 the Niningers moved from Denver to their new location on old Highway 66, not far from Meteor Crater. Conditions were primitive, and there were endless tasks involved in repairing and weather-proofing the building they had acquired, in the moving of their large collection of meteorites (which by that time was one of the major collections in the country), and in the establishment of their museum. After weeks of arduous labor, Nininger was at last able to return to his soil samples.

The soil around Meteor Crater contained a number of extremely interesting components. Relatively large particles, one to two millimeters in diameter, were plentiful in ant hills and appeared to be pellets of sand and soil grains. They gave a positive test for nickel, and seemed to be similar to nickeliferous pellets which had been found at the Haviland crater. Nininger discovered, however, that they had malleable metal cores, to which the sand and soil grains appeared to be cemented through the action of oxides. He interpreted the metallic cores as having been formed by condensation from metallic vapors in the absence of oxygen. If so, they must have been formed in the midst of a large cloud of vaporized metal, from which the air had been blown aside; and if this were the case, other particles must have formed at the edges of such a cloud, where oxygen was present.

Continuing search revealed smaller nickeliferous particles, black in color and less than a millimeter in diameter. These particles were at first difficult to distinguish from ordinary sand, but a critical examination showed that they exhibited surface features characteristic of droplets which have congealed from a melt. They must have formed, in Nininger's opinion, when vapors of the explosive cloud, in contact with the atmosphere, condensed into droplets which were immediately and completely oxidized.

*7.4    All that remains of Nininger's meteorite museum on old Highway 66.*

Still other particles, which Nininger called metallic spheroids, were extremely rich in nickel. These tiny particles of nickel-iron—so small that an ounce would contain 280,000 of them—bore very thin coatings of oxide, which evidently protected them from weathering. They could be explained only as condensation droplets from a cloud of vaporized metal. By early 1948, having examined hundreds of soil samples taken from locations on all sides of the crater, Nininger was convinced that metallic spheroids "were present in the topsoil to an extent that projected to thousands of tons." Here, indeed, was proof of Moulton's explosion theory.[15]

Excited and encouraged, Nininger renewed his search for the absent silica glass. One day in a gravel pit dug into the crater by the state highway department, he noticed tiny "bombs" of what he described as yellow-green-brown slag. A search soon revealed a number of similar small bombs, about the size of grains of wheat. Ground on a carborundum cloth and then held under a pocket lens, they revealed embedded metallic particles as bright as chrome steel. The metallic particles gave a positive test for nickel. Here were condensation products similar to those described by Spencer.

In the autobiographical account of his researches, written in 1972, Nininger described looking in other parts of the crater for similar bombs of glassy-metallic slag. The day after his discovery in the gravel pit, he found on the southeast rim that

> mingled with the rubble of the crater rim were tiny droplets of melted country rock, some as round as bird shot, others pear-shaped, oval, cylindrical with rounded ends, and still others of almost any imaginable shape and ranging from microscopic to the size of walnuts! One could kneel in a single spot and pick up 200 pieces without even moving. Some looked just like those from the gravel pit, but others seemed to be mere volcanic cinders. All, when ground on my carborundum cloth, showed metallic grains.[16]

There was no longer any doubt in Nininger's mind. The meteorite was not buried beneath the crater. He had tangible proof of an explosion.

Several years later, an investigation of the minuscule bits and pieces of meteoritic material scattered through the soil in the neighborhood of the Arizona crater—as pointed out by Nininger—was undertaken by the Smithsonian Astrophysical Observatory. It was estimated that the total mass of meteoritic material in the region was between 10,000 and 15,000 tons.[17]

It is interesting to note, incidentally, that there were some scientists at that time for whom an explosive origin for the Arizona crater had long been an accepted fact. One of them was Lincoln LaPaz, who remarked in 1953 that Nininger's gathering of nickel-iron granules at the Arizona crater was not the epochal discovery that some people claimed. LaPaz pointed out that metallic particles had been discovered there by Barringer and Tilghman in the first decades of the century. He neglected to point out, however, that at that time they were not correctly interpreted.

Metallic granules were indeed mentioned by Barringer, who concluded that the metal must have been in a fused state when it fell to earth. Tilghman, in reporting his meticulous observations at Meteor Crater, mentioned finding tiny globules of meteoric iron, but he suggested that they were part of the luminous tail of the meteor: ". . . the luminous tail in such a case must have consisted of atomized particles of incandescent magnetite . . . a search for this material was made with magnets . . . and it was found that its presence was absolutely universal over the whole locality."[18]

It was not until Spencer's work in the 1930s showed that the volatilization of meteoritic and terrestrial material actually occurred at Wabar and Henbury, along with theoretical calculations of the enormous amounts of heat which must be generated when a large high-velocity meteorite strikes the earth, that it was widely realized that such meteoritic impacts must be explosive. Nininger's discovery of glassy particles containing tiny metallic flakes and spheres was indeed a confirmation of the volatilization of some of the materials involved, and of the explosive origin of the crater.

## CHAPTER 8

# Meteorite Craters
# Known at Mid-Century

DURING THE TWENTY YEARS following the more or less general acceptance of Meteor Crater as an impact scar—from about 1930 to 1950—holes and depressions possibly formed by the falling of meteorites were found in various widely scattered locations. Investigation of them resulted in a gradual increase in the number of authenticated and probable meteorite craters known in the world. Many of them were first observed from the air. In other cases discoveries were sometimes made by people who knew little about meteorites or meteorite craters. Occasionally a hole thought perhaps to be of impact origin was considered just a local curiosity and remained unreported until excitement caused by some meteoritic event suggested that it might be of scientific interest. An example is the Dalgaranga crater in Western Australia, of which no description was published until fifteen years after its discovery.

C. E. Willard, manager of the Dalgaranga sheep station, came upon this crater in 1923. Seven years later, after general interest had been aroused by the sensational daylight fall of the Gundarung meteorite in Western Australia, Willard reported his discovery to E. S. Simpson, a government mineralogist. Simpson did not visit the crater, but eventually, in 1938, he published Willard's description of it. According to Willard, the crater was about 15 feet (4.5 m) deep, with a diameter of about 225 feet (69 m). He stated that large pieces of rock had evidently been ejected from it, as if there had been an explosion, and that in and around it he had collected a hatful of small ragged pieces of iron which had since unfortunately been lost. However, a sample of Dalgaranga iron provided by the current manager of the station proved to be meteoritic. It was found to contain nickel, and an etched surface displayed clear though somewhat distorted Widman-stätten patterns.

In 1958 the Niningers visited the Dalgaranga crater and found it much smaller than they expected. Willard's measurements were evidently incorrect, for according to Nininger the "beautifully shaped, well preserved but obscure little crater" measured only 70 feet (21 m) across and was about 10 feet (3 m) deep. Nininger reported that meteorite fragments around it were not plentiful, but a small supply was collected, and the samples were eventually cut and polished at the meteorite museum in Arizona. They were found to be so unusual and of such remarkable variety that another expedition to the Dalgaranga crater was undertaken the following year. This time its measurements were confirmed, and a slight bilateral symmetry was observed. Use of better equipment than was available on the first visit led to recovery of some 207 small meteoritic specimens.[1]

Another crater in Australia was reported by Joe Webb, a shearer at the Boxhole sheep station in central Australia. In 1937 he took a geologist, C. T. Madigan, to a great hole on the north side of the Plenty River. The hole was almost circular, more than 500 feet (152 m) in diameter, and penetrated the Precambrian rock. Webb suspected that it was a meteorite crater, and his suspicions were confirmed by the discovery of nickel-bearing metallic fragments and iron shale balls similar to those found at Henbury. A later search resulted in the finding of additional meteoritic metal, including an iron mass weighing 181 pounds, which was acquired by the British Museum.[2]

8.1    Wolf Creek crater, in Western Australia, with a diameter of 3,000 feet (915 m).

In 1947, during an aerial survey conducted by an oil company, a large crater was discovered in Western Australia.[3] It was photographed from the air. The results were shown to H. G. Raggatt, director of the Commonwealth Bureau of Mineral Resources, who was impressed by their similarity to photographs of Meteor Crater. Two months later the aircraft crew made a trip to the crater by jeep. The location, in remote desert, was not crossed by scheduled plane flights, nor likely to be visited by prospectors or cattlemen. It lay close to the usually dry bed of Wolf Creek, in the East Kimberly district of Western Australia, and it became known as Wolf Creek crater. A rough track from Hall's Creek south to Ruby Plains and Billiluna cattle stations passed eleven miles from it, but it had escaped notice because from the ground it appeared to be a low, flat-topped hill. It does not seem to have been known to the inhabitants of the surrounding sparsely settled region, though a former constable reported that it had been shown to him in 1935. Although the aborigines were aware of it, they had no legend concerning its origin.

The crater was remarkably symmetrical, about 150 feet (45 m) deep and some 3,000 feet (915 m) in diameter. A rim composed of angular blocks of rock rose 60 to 100 feet (18 to 30 m) above the surrounding desert. Its walls were of hard Precambrian quartzite (also described as sandstone), the layers of which were tilted outward and downward. No volcanic material was found. The conclusion that it was a meteorite crater was confirmed by a group of investigators from the Commonwealth Bureau of Mineral Resources, who discovered weathered meteoritic material similar to the iron shale found at Henbury and the Arizona crater.[4] During a visit to Wolf Creek in 1954, William Cassidy collected more than 1,400 pounds of what he called "more or less completely oxidized meteoritic material."[5]

The first crater to be discovered from the air was Aouelloul (pronounced Ah-way-lewell), in the west central Sahara Desert. It was seen during air travel in 1938 by A. Pourquié, who visited it the same year, and it was noted in 1947 during aerial surveys conducted by the U.S. Air Force in what was then French West Africa. Located on an almost horizontal sandstone plateau, about 250 miles from the Atlantic coast, it was said to have been used by natives of the region as a place of refuge in times of danger and was known to them as the hole of Aouelloul. The crater was a little more than 800 feet (244 m) in diameter and, although partially filled with sand, was about 20 feet (6 m) deep. In 1950 and 1951 black silica glass was discovered in and around the crater, and an analysis at the British Museum showed that it contained small amounts of nickel. This supported suggestions that the great hole was the result of meteoritic impact.[6] Impact origin was later confirmed by the discovery of nickel-iron spherules in the Aouelloul glass.[7]

Three other craters, the Richat dome, Tenoumer, and Temimichat, were discovered north of Aouelloul, almost perfectly aligned in a northeast direction over a distance of some 360 miles. There has been much speculation about their origin. The Richat dome, a circular topographic depression about 18 miles (30 km) in diameter, was described by several investigators. In 1964 A. Cailleux and his associates claimed to have found evidence that it was of impact origin. Later, in 1969, after a careful examination, a group including R. S. Dietz was unable to confirm this claim.[8] In their opinion, neither the Richat dome nor Semsiyat, a much smaller structure nearby, was the result of impact. Evidence that Tenoumer was an impact structure was reported in 1970.[9] Because of the remarkable alignment of the three craters—Aouelloul, Tenoumer, and Temimichat—and their apparently similar ages, it has been suggested that Temimichat, like the others, may also be meteoritic, and that the three structures may be the result of a triple cratering event.

The fact that the Aouelloul crater was within a few miles of the general location in which the Chinguetti meteorite had been discovered some years earlier was pointed out in 1954 by L. and J. LaPaz. The date of discovery of the Chinguetti meteorite seems uncertain. According to T. Monod, it was found in 1916, but later dates have been suggested. M. H. Hey gives 1920 as the year of its discovery. Monsieur Ripert, a government official, reported that it was known to natives of the region, who called it "the stone that fell from heaven" and seemed anxious to conceal its existence from Europeans. Ripert, however, persuaded a native leader, who at first denied any knowledge of it, to show it to him. With some reluctance, the native guided him to the spot, stipulating that he might carry along no compass or materials which would permit him to make notes or measurements. Whether the subsequent death of the native, apparently by poisoning, was related to this incident is not known. In any case, in the midst of the Adrar desert, Ripert reported finding a metallic fragment lying upon an enormous metallic mass (said to be a source of iron for native blacksmiths) which he judged to be about 325 feet long and 130 feet high. This enormous mass of metal, according to his description, had the form of a "compact, unfractured parallelopiped."[10] The visible part of it was vertical, standing like a cliff-face above a wind-excavated hollow; and parts of it had been polished by wind-blown sand until they shone like mirrors. Because much of it was buried by sand, the third dimension of the mass could not be determined. On its summit the wind had carved out tiny metallic needles of malleable iron, which Ripert was unable to remove or even to loosen.

Ripert eventually turned the metallic fragment over to a friend, H. Hubert, "A Doctor of Science at Daker who," he remarked, "did not seem to attach any great importance to its existence."[11] Hubert, in turn, passed it on to A. Lacroix, along with Ripert's account of its discovery. A detailed description of it and of the manner in which it was found was published by Lacroix in 1924; and this account, which contained analyses by R. Raoult showing that the metal in the fragment was largely nickel-iron, was included in translation in the LaPaz paper.

After describing his careful mineralogical examination of the fragment of the Chinguetti meteorite, which weighed 4.5 kilograms and had the form of a "flat parallelopiped," Lacroix concluded that "the Chinguetti meteorite constitutes a type of siderite [e.g., iron meteorite] differing from all those known up to the present time."[12] But the large metallic mass, upon which the fragment was found and which many investigators considered even more interesting than the

8.2 Aerial photograph of the Richat dome, in northwest Africa, taken at a height of 165 nautical miles. It is a circular topographic depression probably created by erosional processes during a long history of slow vertical uplift.

**8.3**   *Aerial photograph of the New Quebec (formerly Chubb) crater, taken at a height of 20,000 miles (about 6 kilometers). No. T193R-87.*

peculiarities of the fragment itself, was never located again. Several searches were undertaken; but the Adrar region, an upland desert area in the western Sahara about 350 miles north of Dakar, and described in the LaPaz paper as "unmapped, wind-blasted, sand-shrouded, and torrent-swept," with constantly changing topography, was not easily explored. The great mass of iron, already partly covered, may have disappeared beneath encroaching sand dunes. Or, as some skeptical scientists did not hesitate to suggest, it may never have existed; the whole thing may have been the result of an imagination overheated by the Saharan sun. Others argued that such a huge piece of metal could not have arrived from space intact. L. and J. LaPaz, however, suggested that the close proximity of the Aouelloul crater to the locality where the great meteorite was said to lie might have considerable significance. They noted that the original data concerning the meteorite were factually presented and internally consistent; and they referred to calculations by R. N. Thomas and F. L. Whipple, who claimed that meteorites very much larger than Hoba West or Ahnighito might be capable of surviving collisions with the earth. But the location and even the existence of the metallic mass upon which the Chinguetti meteorite was found remain unsolved mysteries of the desert.

During a routine weather flight in 1943, a small lake was photographed from a U.S. Army Air Force plane. The lake was remarkably round and later appeared in photographs taken by the Royal Canadian Air Force in 1946. It was located in the Ungava district of northern Quebec near the headwaters of the Povungnituk River, and it soon appeared on maps of the region bearing the label "crater."

In 1946 W. K. Carr, of the Royal Canadian Air Force, was in charge of transporting personnel connected with a program of aerial mapping in northern Quebec. Flying over uncharted country presents many challenges and, as he remarked, a pilot quickly learns to remember outstanding features of the terrain. The Ungava crater was such a feature. He reported that one day in 1946, after delivering a geodetic party to a location some 300 miles north of the base camp, he encountered bad weather on the return trip. He took refuge on the round lake, which happened to be near, and remained there until the squall had passed. He described the crater, when viewed from the inside, as appearing to be the result of a big "splash" of some sort. It did not appear to be volcanic; and the lake, unlike other lakes of the region, was enclosed by a high, steep rim of tumbled rock. Its possible origin became a subject of discussion at the base camp, and it continued to be a landmark for pilots, as the rim could be seen for a distance of 20 to 30 miles.

When asked to interpret the crater, Y. O. Fortier, of the Geological Survey of Canada, noted its similarity to the Arizona crater and remarked that a picture of the Ungava crater had already been selected as an example of a meteorite crater. Another comment appeared in a 1949 report on the Ungava region by G. V. and M. C. V. Douglas, from which the following was quoted by P. M. Millman:

> The crater (16°18′N. Lat., 73°40′W. Long.) . . . is thought to be meteoric in origin. However, this crater has never been examined on the ground, and the theory is based on a study of aerial photographs. If the crater is meteoric . . . it [i.e., the meteorite] fell nearly vertically, for the elevations on all sides are about the same . . . the circular section and the very steep internal slopes also indicate meteoric origin.[13]

A prospector, F. W. Chubb, chanced to hear of the Ungava crater and thought it might perhaps be the upper end of a volcanic vent, or diatreme. In Africa, diamonds have been found in diatremes, and Chubb was aware of the fact that diamonds had occasionally been found in glacial gravels which originated in northern Quebec. In February 1950 Chubb pointed out the existence of the crater to V. B. Meen, director of the Royal Museum of Geology and Mining in Toronto. Meen's interest was immediately aroused. By mid-July funds had been procured and preparations for an expedition were completed; and the two men, along with a photographer, R. C. Hermes, and a correspondent from the Toronto *Globe and Mail*, set out to visit the crater, a flight of some 1,700 miles from Toronto. (The aircraft and its pilot and flight engineer were supplied by the *Globe and Mail*, and it was largely due to energetic efforts on the part of the newspaper correspondent, K. W. McTaggart, that funds from several anonymous donors were made available.)

The round lake was in a rocky, subarctic region of the Ungava district of northern Quebec, some 290 miles northwest of Fort Chimo. As the lake was almost covered with floating ice, the amphibian aircraft was set down on a nearby ice-free lake, for which Meen suggested the name Museum Lake. The round lake itself he proposed to call Chubb Crater Lake. Camp was established in a small area of boulder-free sand in what he described as a most barren, forbidding region, with boulders covering the ground as far as the eye could see. The crater lake was estimated to be about 10,000 feet (3,050 m) in diameter. It was edged by a steep rim which rose at an angle of 45 degrees to heights of more than 500 feet (150 m) above the surrounding terrain. This ruled out the possibility that it might be a sinkhole, or caused by the melting of a pocket of glacial ice. No volcanic rocks were found in the vicinity, nor was there any indication of a diatreme. Meen became convinced that it was the result of meteoritic impact; and if so, it was the largest meteorite crater then known.

The rim was composed of great blocks of granite representing in Meen's opinion "a mass of granite bedrock which has been fractured by a tremendous explosion and lifted bodily to its present position."[14] More or less concentric ridges of bedrock were observed some distance outside the rim. Meen remarked that they might "indicate shock compression ripples similar to those produced in calm water when a stone is dropped in."[15] No evidence of glaciation was observed. Meen suggested a post-glacial age for the crater of less than 15,000 years.

The following year (1952) additional observations were made during an expedition sponsored by the National Geographic Society and the Royal Ontario Museum. This time the diameter of the crater was determined to be about 11,500 feet (about 3,500 m). Although the search for meteorite fragments was still unsuccessful, the investigators were encouraged by the discovery of a magnetic anomaly on the northeast rim, possibly indicating the presence of buried meteoritic material. They pointed out later, however, that "no anomalies have yet been found which could not be satisfactorily explained by slight concentrations of magnetite in some bands of the country rock."[16]

In 1954 many indications of glaciation at the Chubb crater— such as rocks with glacially polished surfaces, striations in rocks on and near the rim, and delicately balanced boulders which could have been emplaced by a melting glacier but could hardly have been dropped into their precarious positions by an explosion—were pointed out by J. M. Harrison. After a reevaluation, Meen agreed that glaciation must have taken place after formation of the crater. The movement of masses of glacial ice over the crater lake, which must have been solidly frozen for a very long time, offered a possible explanation for the absence of meteorite fragments. So much other evidence seemed to indicate an impact origin for the Chubb crater that it was accepted as a probable meteorite crater by many investigators. In 1954 its name was officially changed to the New Quebec crater, and the lake within it was given the name of Crater Lake. The name of Museum Lake was changed to Lac Laflamme.[17]

Several other craters were discovered close to the same time. Soon after the discovery of the New Quebec crater in Canada, R. Karpoff, of Paris, investigated an interesting circular feature some 240 miles south-southeast of Algiers. Karpoff reported that it was a

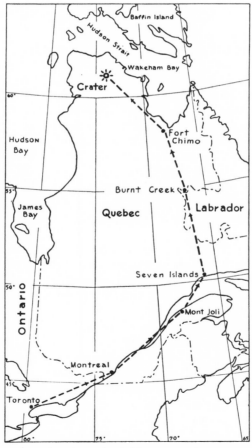

*8.4    Location of the New Quebec (formerly Chubb) crater, and route taken by the Royal Museum – Globe and Mail expedition in July 1950.*

"magnificent circular bowl of imposing dimensions," located 5 miles directly south of a few buildings known as Talemzane. A flight over it was arranged early in 1952, and a number of aerial photographs were taken. Karpoff made several exploratory visits to the structure and proposed to call it the crater of Talemzane.

The crater had steep inner slopes of light-colored limestone and a diameter of more than 5,000 feet (1,525 m). Around its edge the uppermost limestone strata of the region, which elsewhere appeared almost horizontal, were steeply upturned, and the rim contained great limestone blocks piled pell-mell one upon another. It was noted, as a possible indication of great age, that these piles of rock had been so planed by erosion that they could be mistaken for a natural level surface; in some places, ancient crevices were filled with breccia. Several investigators took part in an unsuccessful search for meteoritic material. In spite of its absence they came to the eventual conclusion that the great hole had been formed by the impact and explosion of a meteorite.

In France in 1952 several craters were observed in aerial photographs of the Hérault region, and a meteoritic origin was suspected.[18] However, the absence of raised rims and uplifted strata was noted in 1964 by C. S. Beals, in whose opinion they were not meteorite craters. Also in 1952, a note in the *Irish Astronomical Journal* drew attention to two craters in the highlands of east Pamir, just north of the border of India. Some 250 years ago, according to local legend, two large meteorites fell near the town of Murgab, blasting holes in the ground. They became known as the Murgab craters. Locally the site is called Chaglgan Toushtou, which means "the place where lightning fell." The larger crater was described by H. D. Klavins in 1926. The second one was discovered in 1951 by Russian investigators.[19]

Although the discovery of these various craters drew the attention of many scientists, there was no great interest on the part of the general public. In 1947, however, a spectacular meteorite shower occurred in eastern Siberia. Reports of it appeared in newspapers all over the world.

On the morning of February 12, 1947, a dazzlingly brilliant bolide streaked across a clear sky, emitting sparks and leaving a trail of dust which remained visible for several hours. Metallic fragments were spread over an elliptical area of almost a square mile on the western spurs of the Sikhote-Aline mountain range north of Vladivostok. The fall of the meteorite, visible over a large area, was seen by several hundred people, and its disappearance behind the hills somewhere at the western edge of the Sikhote-Aline was followed by reverberating detonations. Observers agreed that its passage across the sky lasted only a few seconds. Later, information obtained through the questioning of witnesses enabled investigators to plot its atmospheric trajectory. Several people reported that before it disappeared it split up into pieces which fell vertically to the ground like lumps of fire. A dramatic picture of the event, painted by P. I. Medvedjev, now hangs in the Mineralogical Museum of the Academy of Sciences in Moscow.

The location of the fall was discovered from the air, revealed by fresh craters in which yellow-brown clay and rocks stood out clearly against the snow. A group of investigators made their way to it, not

*8.5   Fall of the Sikhote-Aline iron meteorite shower.*

without difficulty, for the spot was deep in almost impenetrable forest. Close to the site many trees were uprooted, tree-tops and branches were broken off, and metallic fragments had penetrated tree trunks. During the next four years, months were devoted to the investigation of the many small craters and the collecting of meteoritic fragments. Some of the larger metallic pieces were discovered in small craters, and numerous specimens were found lying on the ground. E. L. Krinov described collecting them soon after the snow had melted and before new grass and leaves had appeared, when "the meteorite fragments, violet-blue in color, like ripe plums, were conspicuous against the yellowish background of the dead leaves."[20]

Some 200 craters were located, and the diameter of the largest was less than 100 feet (30 m). In connection with the excavation of one of the larger craters, 75 feet (23 m) in diameter, Krinov remarked that "a meteorite of considerable mass, possibly several tons in weight, formed this crater and at the same time split up into numerous, mainly tiny, fragments that were scattered over the slopes and in the vicinity of the crater."[21]

The largest meteorite specimen, which weighed 1,745 kilograms, was found in what was described as a medium-sized crater. Many of the craters were very small; in fact, seventy-eight of them were less than two feet across. There was a large quantity of meteoritic dust, but no indication of an explosion nor of great heat. No disturbances were reported at the nearest seismic station at Vladivostok; evidently the entire force of the fall was much less than that of the Tunguska event of 1908. Investigations were resumed several years later (1967 to 1970), and a great deal of additional meteoritic material was found.

In 1933, as mentioned earlier, five "more or less certain examples of known meteorite craters" were listed by L. J. Spencer. These were Meteor Crater, Arizona; Odessa, Texas; Henbury, Australia; Wabar, Arabia; and Campo del Cielo, Argentina. Four other craters of possible meteoritic origin were known. Two of them, Kaalijärv and Tunguska, were groups of craters without meteoritic material, and were therefore considered unproved. The other two, Lake Bosumtwi and the crater in Persian Baluchistan, single craters where no meteoritic material had been found, he considered doubtful. By 1950 the recognized number of "certain examples" of meteorite craters had more than doubled. A list of the authenticated meteorite craters of the world, totaling eleven, was compiled by F. C. Leonard in 1946. It contained the first seven listed by Spencer, omitted the doubtful ones (Lake Bosumtwi and Persian Baluchistan), and added three more: Haviland, Boxhole, and Dalgaranga. Leonard also included Mount Darwin, in Tasmania, where silica glass had been discovered and where it was thought that a meteorite crater now obliterated by erosion had once existed. A list drawn up in 1954 by C. T. Hardy did not include Mount Darwin but added Sikhote-Aline and Wolf Creek.

In addition, a number of "probable" or "possible" meteorite craters were known; and although no meteoritic material had been found near them, several investigators were firmly convinced that some of them were impact craters. Among the best known of the "probable" examples were Aouelloul; Lake Bosumtwi; Lonar Lake, a solitary circular depression in the basalt of the Deccan Plateau in India (which, incidentally, was referred to by G. K. Gilbert in 1896 as an example of a steam explosion); Talemzane; and two African craters, the Pretoria Salt Pan and the Vredefort dome—although this last one, some fifty miles (80 km) across, appeared because of its great size to be quite different from the others. The doubtful one in Persia did not appear on any of the lists of authenticated meteorite craters.

At mid-century, the following twelve locations were generally accepted as sites of meteoritic impact. Meteoritic material had been discovered at all of them except Tunguska.

| | |
|---|---|
| Arabia: | Wabar |
| Argentina: | Campo del Cielo |
| Australia: | Boxhole |
| | Dalgaranga |
| | Henbury |
| | Wolf Creek |
| Estonia: | Kaalijärv |
| Siberia: | Sikhote-Aline |
| | Tunguska |
| United States: | Haviland |
| | Meteor Crater |
| | Odessa |

Prominent among possible sites of meteoritic impact were:

| | |
|---|---|
| Africa: | Aouelloul |
| | Lake Bosumtwi |
| | Pretoria Salt Pan |
| | Talemzane |
| | Vredefort Dome |
| Canada: | New Quebec |
| India: | Lonar Lake |

It is interesting to see that L. J. Spencer, writing in 1937, did not add the Haviland crater to his list of "really certain examples of meteorite craters"—much to the chagrin of Nininger, who, after all, had found some 1,200 pounds of meteoritic material there in 1933. Actually, Spencer's omission of Haviland is one more illustration of the difficulties which were encountered in achieving an understanding of meteorite craters.

In 1933, as we have seen, Spencer pointed out that meteorite craters "are not merely dents or holes made just by the projectile force of the meteorite, as hitherto supposed. They appear rather to be explosion craters due to the sudden vaporization of part of the material both of the meteorite and of the earth, in the intense heat developed by the impact."[22] He found confirmation for this view in Gifford's calculations of the immense amounts of energy that must be released on impact by large high-velocity meteorites. He pointed out that such impacts must be explosive, creating round holes regardless of the angle from which the meteorite approached. Assuming that most lunar craters were caused by meteoritic impact, this offered an explanation for the scarcity of elliptical craters on the moon. In addition, it explained the presence of silica glass at the sites of meteoritic impact, and it also accounted for the absence of meteorite fragments within terrestrial meteorite craters, as such fragments would be "backfired" out of the crater by the explosion.

With these considerations in mind, Spencer developed what Nininger described (with some justification) as an arbitrary definition of the term *meteorite crater*—a definition into which the Haviland crater did not fit. The Haviland crater was small. It was elliptical. It contained no silica glass or other indications of great heat. And a large quantity of meteoritic material had been found *within* it.

Only gradually was it realized that because meteorites travel at various velocities and may be of almost any size, the energy released on impact ($\frac{1}{2}mv^2$) must vary widely from one event to another. In the case of a large high-velocity meteorite the energy released on impact is enormous. In the case of smaller, slower bodies, however, whose lower speed may be still further reduced by passage through the atmosphere, the energy released may not be large enough to cause an explosion. Such an impact must therefore be percussive rather than explosive, and such meteorites may be found on the surface of the ground. An example is the Hoba West meteorite, which is thought originally to have weighed eighty tons or more. It was found embedded in soft white limestone (described as calcareous tufa, probably spring-deposited after the fall of the meteorite), with its upper surface partly exposed, and with no trace of a crater. In other cases, relatively small pits are formed, usually funnel-shaped, with an elliptical outline at the ground surface, and frequently containing meteoritic material. Such craters are called "percussion pits" or "penetration funnels." The Haviland crater is an example of this type—a true meteorite crater, formed by percussive rather than explosive impact. The Henbury group includes both explosive and percussive craters, and this distinction was recognized by Reinvaldt in his investigation of the Estonian craters. In the Sikhote-Aline event there was apparently no evidence of an explosion. It seems probable, in fact, that all of the many craters resulting from that spectacular fall were percussive.

## CHAPTER 9

# The Craters of the Moon

ALTHOUGH DURING THE EARLY decades of the twentieth century several holes on the surface of the earth were positively identified as meteorite craters, it was generally thought that the craters on the moon indicated some sort of volcanism. Striking differences between the lunar forms and terrestrial volcanoes were pointed out in 1892 by G. K. Gilbert, but the significance of the differences was not widely appreciated. One argument against a meteoritic origin for lunar craters—the fact that few of them were elliptical—was answered, as we have seen, when it was realized that the explosive impact of a large high-velocity meteorite must result in a round hole. But in the opinion of many investigators, the fact that so few meteorite craters exist on the earth presented an insurmountable objection to the idea that lunar craters were meteoritic: for if the immense number of craters on the moon had indeed been caused by meteoritic impact, the nearby earth would be similarly scarred. And quite obviously it was not.

In the opinion of most nineteenth-century scientists, lunar craters were caused by some kind of volcanism—an opinion supported and strengthened by an influential book, *Der Mond . . .*, by W. Beer and J. H. Mädler, which appeared in 1837. These authors named, mapped, and measured an enormous number of lunar craters, which they considered to be extinct volcanoes. A map of the moon showing more than 32,000 craters and many other features was published in 1878 by Julius Schmidt. So meticulous and painstaking was the work of these and other investigators that many people considered the problems presented by the moon to be solved. Nevertheless, other opinions emerged from time to time. According to Jack Green, a meteoritic origin for lunar craters was suggested as early as 1802 by the von Bieberstein brothers,[1] when the existence of meteorites themselves was still a matter of discussion. The best known, perhaps, of several similar early suggestions is attributed to F. von P. Gruithuisen. In 1873 the British astronomer R. A. Proctor wrote:

> It may seem indeed at a first view, too wild and fanciful an idea to suggest that the multitudinous craters on the moon, and especially the smaller craters revealed in countless numbers when telescopes of high power are employed, have been caused by the plash of meteoric rain, — and I should certainly not care to maintain that as the true theory of their origin; yet it must be remembered that no plausible theory has yet been urged respecting this remarkable feature of the moon's surface. . . .[2]

Proctor evidently decided that his idea was indeed too wild and fanciful, for he later discarded it.

At the beginning of the twentieth century, most interested scientists—mainly astronomers and geologists—supported theories involving volcanism as the cause of lunar craters, though there were some who took no stand in the controversy, preferring to acknowledge their ignorance. Hooke's suggestion of bubbles on a molten surface (a process of "degassing") was revived by some investigators. Some, in agreement with J. D. Dana, compared lunar craters with the shield volcanoes of Hawaii. Others compared them to terrestrial maars. Still others explained certain lunar formations as resulting from laccolithic processes similar to those described by Gilbert in his study of the Henry Mountains. Among suggestions which did not involve volcanism was the tidal theory, which attributed lunar craters to the action of enormous tides that periodically caused fluid rock from the moon's interior to well out on the lunar surface. It was suggested that when the tide subsided and the fluid sank back into the moon, a circular residue of cooled and congealed rock would mark the edge of the area that had been covered. After repeated upwellings, the process would result in the formation of circular crater-like structures. Various other non-volcanic theories involving such things as ice on the surface of the moon, and the influence of sunspots—and even coral reefs—were short-lived.

Some investigators supported an impact origin for only the very large lunar scars—the maria. Others gave unqualified support to the impact hypothesis. T. J. J. See, for example, announced in 1910 that he was "satisfied that the impact theory is correct." After listing some twenty-four reasons for his opinion, he concluded:

> we may dismiss the old volcanic theory once and for all as false and misleading; and may look upon our satellite as a battered

planet which presents to us the most lasting and convincing evidence of the processes of capture and accretion by which the heavenly bodies are formed.[3]

Another supporter was D. P. Beard, who referred in 1917 to a geologically remote period when "the moon was bombarded and gashed by colossal meteoric missiles from the sky."[4] And it has sometimes been said that the people most firmly convinced that the meteoritic theory was correct were the "amateurs" who in many cases were specialists in other fields, as illustrated in a letter written in 1914 to D. M. Barringer by the versatile Elihu Thomson:

> I am glad that you have developed in your paper[5] the analogies between the lunar craters and the Arizona Meteor Crater, for I really believe that there is a great deal of need of people understanding the significance of the forms taken by the lunar craters. There has been too much dogmatism on volcanic origin, etc. which could never be reasoned out satisfactorily. I have never read any theory of the volcanic origin of lunar craters which at all satisfied me that such craters could be produced in that way. . . . I think the rain of meteors must have continued for many thousands, if not millions of years, and it is probably not over yet. . . .[6]

In the paper referred to above, Barringer stated his firm belief that most of the craters on the moon's surface had been caused by impact. This opinion was shared by most of the people connected with the work at the Arizona crater.

Among the supporters of meteoritic impact as the cause of lunar craters was the German meteorologist Alfred Wegener, who gave careful consideration to the four theories current among German scientists at that time. His paper, written in German in 1921, is now considered an excellent account; a 1975 translation was supported by a National Aeronautical and Space Administration grant. It seems, however, to have had rather little influence when it first appeared. It may be that Wegener was considered suspect because of his ideas about continental drift, which at that time were generally regarded as anything from heretical to absurd.

With respect to the "bubble hypothesis" Wegener remarked that anyone who explained the origin of giant lunar features hundreds of kilometers in diameter with broken bubbles would be guilty of "a fallacy comparable to one who wants to explain the flotation of transatlantic ships with surface tension." He disposed of the tidal theory as requiring a firm, solid lunar crust which would be unaffected by tidal waves in the underlying material, claiming that because of isostasy (or pressure balance) the tidal hypothesis was "as impossible as to try to measure the tide and ebb from icebergs floating in the North Sea." In a discussion of the volcanic theory, he compared lunar and terrestrial forms, much as Gilbert had done some thirty years earlier, and concluded that "the forms are fundamentally different; therefore the origins also should be different." Typical lunar craters, he said, can best be interpreted as impact craters; and he went on to deduce from the available facts that the impacting bodies belonged to the solar system and that the moon was built through a process of accretion. Although he thought it highly improbable that Arizona's meteorite crater was the only impact scar on the earth, he did not offer a convincing argument to explain the lack of other terrestrial evidences of impact. "The important thing," he said, "is that the absence of large impact craters on the Earth can in no way be used against the hypothesis of lunar impacts."[7]

A decade later, an implicit explanation of the perplexing absence of terrestrial impact craters appeared in the little-noticed work of Boon and Albritton. And L. J. Spencer, after reading Gifford's "overlooked" paper, felt that the absence of elliptical craters on the moon no longer presented a problem. As for the absence of ancient meteoritic materials on the earth, considered by some people to be the only infallible indication of meteoritic impact, "these," he said, "will be destroyed by weathering and all traces of them obliterated by denudation long before the next geological period." Traces of ancient impact events, such as craters, would also have disappeared. "The argument based on the absence of pitting on the earth's surface," he said, "could never have come from a geologist."[8]

It is a curious fact that when some stubborn problem obstructs progress along a particular line of investigation, it sometimes happens that solutions are achieved, independently and more or less simultaneously, by quite unconnected investigators. A coincidence of this sort occurred in the 1940s.

In 1942 and 1943, two papers in support of the meteoritic origin of lunar craters were published—not without difficulty—by R. B. Baldwin. He later reported that his first paper was rejected by several well-known scientific journals before it finally appeared in *Popular Astronomy*. In these papers Baldwin suggested that certain grooves radial to some of the large lunar structures presented an argument for meteoritic impact. He pointed out later that he was at the time unaware that similar arguments had been made by Gilbert. In 1946 R. S. Dietz published a paper on the meteoritic origin of lunar craters. Although the far side of the moon was sometimes referred to as the greatest unsolved lunar mystery, "this mystery," Dietz remarked, "is probably a poor second as compared with the enigma of the origin of surface features on 'this side' of the moon."[9] Were those features the result of volcanism, as maintained by most investigators, or of meteoritic impact? Or did both forces operate on the lunar surface? Dietz gave various arguments in support of meteoritic impact. Although in some cases similar interpretations had appeared in the work of earlier investigators, his paper was the result of independent observation and study, and it was the most comprehensive and exhaustive account of the lunar surface since Gilbert's time. It was not well received by other scientists, however, most of whom supported volcanic theories.

A slim but influential volume entitled *The Face of the Moon,* by R. B. Baldwin, appeared in 1949. Many years later, in an account of his preparation for this book, Baldwin remarked that he had found plenty of opinions expressed in various publications, but "little real evidence pertinent to the overall problem of crater origin." But, he went on,

> there was one real exception, and I have to say that I missed it entirely while collecting information for *The Face of the Moon.* It was Robert Dietz's monumental paper (Dietz, 1946) in the *Journal of Geology* entitled *Meteoritic Impact Origin of the Moon's Surface Features.*[10]

In spite of the fact that their investigations were entirely independent of one another, Dietz and Baldwin arrived at many of the same conclusions.

It was observed by many scientists, including both Barringer and Tilghman, that a more or less constant depth-to-width relation exists for a given type of crater. In *The Face of the Moon* Baldwin came to grips with this problem, for it was clear to him that in order to determine the origin of the lunar craters it was necessary to discover whether any relationship existed between them. "If definite relationships are found," he remarked, "one may reasonably expect that these structures were each formed by variant applications of the same force."[11]

Although earlier investigators who charted and studied the moon did not have the advantages of modern instruments, they had, as Baldwin pointed out, the qualities of patience and perseverance. With relatively primitive equipment they measured the depth and diameter, and the height and slope of the enclosing rims, of a large number of lunar craters. Baldwin drew upon the findings of various earlier scientists, particularly those of J. H. Schröter, H. Ebert, and T. L. MacDonald. He pointed out that Schröter, in work published at the end of the eighteenth century, reached the conclusion (known as Schröter's rule) that for each crater the part of the material above the surface is approximately equal to the volume of the depression below the surface; and that Ebert, in 1889, observed that a diameter-to-depth relationship existed among the lunar craters which he had measured. In 1931 MacDonald concluded that small deviations from Ebert's diameter-to-depth relationship were probably related to local conditions on the lunar surface. Baldwin, using measurements obtained by his predecessors along with his own, found in agreement with Ebert that consistent diameter-to-depth relationships could indeed be observed in lunar craters, and also that a slight change in diameter-to-depth proportions occurred in a progression from very small to very large craters. And he showed that this relationship was not affected by local conditions on the lunar surface.

Baldwin then made a remarkable observation. He examined the measurements of bomb craters and of terrestrial meteorite craters, and he found that exactly the same diameter-to-depth relationship observed in lunar craters also existed in these explosion craters on the earth. In his opinion, while some crater-like structures of an entirely different form and probably of volcanic origin could be observed on the moon, the preponderance of lunar craters—the circular, rimmed depressions that covered so much of the moon's surface—were explosion craters. On the earth, naturally occurring explosion craters were caused by the impact of meteorites. On the lunar surface, too, they must be meteorite craters. Or, in Baldwin's words, "The case for the explosive origin of the moon's craters is unassailable. The probability is very great that the explosions were caused by the impact and sudden halting of large meteorites."[12] And, incidentally, one of the arguments in support of a meteoritic origin for the New Quebec (Chubb) crater arose from the fact that when due allowance was made for the effect of glaciation on the height of the rim, the diameter-to-depth proportions of the structure were in close agreement with those given by Baldwin for explosion craters.

Baldwin, moreover, had an answer to that recurring question: if the craters on the moon were caused by the impact of meteorites, why is the surface of the earth not similarly cratered? In his opinion, the earth had not escaped meteoritic bombardment, but its effects had been obliterated by processes of erosion—processes which in the absence of a lunar atmosphere do not occur in a similar manner on the moon. "Written in the Book of Geology in still obscure characters," Baldwin remarked, "are the records of hundreds of thousands of collisions of the earth and extraterrestrial bodies."[13]

Rather than ending the controversy over the origin of lunar craters, however, the work of Dietz and Baldwin drew wider attention to it. And the controversy continued.

## CHAPTER 10

# Coesite and Shatter Cones

THERE WAS LITTLE IMMEDIATE interest in the suggestion made by
J. D. Boon and C. C. Albritton in 1936 that cryptovolcanic structures,
rather than indicating some sort of muffled volcanism, were instead
the eroded remnants of impact scars. Attention was drawn to it in
1947, however, by the widely respected geologist Reginald A. Daly.

In 1922, as a member of the Shaler Memorial Expedition, Daly
visited South Africa, which he described as "crowded with geological
wonders." One of the most startling of these was the vast ring structure of the Vredefort Mountain Land in the Orange Free State. Daly
and a colleague, Charles Palache, both of Harvard University, and
F. E. Wright, of the Geological Laboratory of the Carnegie Institution
(Washington), were guided through this complex and puzzling region
by two outstanding South African geologists, A. L. Hall and G. A. F.
Molengraaff. Exploration of the region disclosed geological problems
which, as Daly noted in 1947, had not been satisfactorily solved by
any of the various explanations offered by a number of geologists. All
these interpretations were based upon recognized terrestrial forces.
Perhaps, Daly suggested, it would be "advisable to take seriously the
published suggestion of Boon and Albritton that the structure may
have been developed by the infall of a speedy meteorite of planetoid or
asteroid dimensions."[1] Daly went on to quote S. J. Shand, of the
University of Stellenbosch, who attempted in 1916 to explain the
characteristics of some peculiar rocks found in the Vredefort region,
and who almost apologetically suggested a hypothesis for their origin
which "takes us away from known causes altogether." His suggestion
involved not meteoritic impact, but some sort of volcanic explosion.
He remarked nevertheless that "when one is dealing with an extraor-

dinary phenomenon, no possibility is too extraordinary to be worthy of consideration."[2]

Daly was a devoted and energetic geologist. Often attacking problems with unconventional and imaginative ideas, he made contributions to almost every branch of geological science. As presented in 1947, however, his view of the possibly meteoritic origin of the Vredefort structure was somewhat clouded by his belief that all meteorite craters then known were the result, wholly or to a large extent, of percussion. He referred to L. J. Spencer's paper, *Meteorite Craters as Topographic Features of the Earth's Surface,* published in 1933, in which Spencer remarked that nothing was then known of the mechanics of the formation of such craters. Pointing out that Spencer recorded evidence of volatilization of both silica and nickel-iron at Wabar, Daly remarked that "the explosive energy of such volatilized materials escapes estimation"[3]; but he noted that no evidence of such volatilization had been reported by observers at the Arizona Meteor Crater. He was, of course, unaware of Nininger's investigations, which were going on at the same time; but he also seems to have been unaware of Spencer's later paper, *Meteorites and the Craters of the Moon* (1937), which, as we have seen, contained a vivid description of the explosive impact of large, fast-moving meteorites. This unfortunate oversight, incidentally, is an example of the lack of communication which almost inevitably occurs now and then among scattered investigators. It was particularly ironical in Daly's case, for he was extremely well informed and broadly interested.

Even without the affirmation which might have been derived from Spencer's 1937 paper, Daly's review of various opinions concerning the Vredefort ring structure led him to conclude that Boon and Albritton's suggestion might well be correct. He remarked that the apparent absence of volatilization at the Canyon Diablo crater had been explained by some investigators as due to its having been formed by a relatively slow-moving meteorite or swarm of meteorites (an explanation later shown to have little validity). As the only known evidence of great heat had been found in small amounts of fused sandstone, most of the energy produced by the impact must have been expended in ejection and pulverization of the rocks. He noted that in the case of the enormous Vredefort structure, the impinging meteorite must have been many times larger. For this reason, it would have been less affected by air resistance and must therefore have had a much greater velocity. The energy released by a huge and rapidly-moving meteorite in its collision with the earth must have been so great that some volatilization of terrestrial and meteoritic materials undoubtedly occurred. The impact, Daly concluded, must therefore have been explosive as well as percussive.

As we have seen, terrestrial maars were known to have formed as a result of volcanic explosions accompanied by the extrusion of little volcanic material. By many investigators they were considered to be the earthly forms which, in spite of their lack of radial symmetry, most nearly resembled lunar craters, and which therefore suggested a volcanic origin for the lunar forms. Meteor Crater in Arizona, on the other hand, was considered by some to be similar in origin and form to a small lunar pit, even though it lacked the central peak characteristic of many lunar craters. The existence of a third group of terrestrial structures having many similarities to lunar craters, and possibly a similar origin, was pointed out by R. S. Dietz. These were the circular areas of intensely disturbed, uplifted, and shattered rocks known as cryptovolcanic structures. "These features," Dietz remarked, "are sufficiently alike to suggest that they were all formed in a similar manner and they differ sufficiently from other known geological structures to indicate that they form a specific structural type."[4]

It seemed to be generally agreed that these structures had been formed by violent natural explosions; but Bucher's suggestion that such explosions could be caused by gases derived from subterranean lava was, in Dietz's opinion, highly questionable. If this were the case, he believed the explosions would have been accompanied by extrusions of volcanic materials, and that escaping gases would have caused some chemical alteration of the rocks. In addition, the events would probably have taken place in volcanic regions. But according to Bucher's definition, cryptovolcanic structures contained no volcanic materials and occurred where there was no evidence of previous volcanic activity.

In agreement with Boon and Albritton, Dietz concluded that these features had been formed by the impact and explosion of large, high-velocity meteorites, and that the so-called cryptovolcanic formations were the "root" structures which remained after the explosion craters had been removed by erosion. He noted that the impact theory gained strength from the fact that the Sierra Madera structure in Texas had been drilled up to a depth of more than two miles without revealing the presence of a core of igneous rock such as would have indicated an intrusion of subterranean lava in the geologic past. In fact, with increasing depth, the structure seemed to die out. Dietz also made an interesting observation concerning the deformations called shatter cones, which were found in the rocks of some cryptovolcanic structures (and which were later shown to be indicative of explosive shock). He observed that shatter cones found in rocks of the Kentland disturbance were oriented in a manner which indicated that the enormous compressional forces which created them had come from *above.* Cryptovolcanic pressures originating in a deep-seated explosion could have come only from below.[5] Nevertheless, Dietz admitted that the source of the explosive force was problematical. He suggested later that until the manner in which the so-called cryptovolcanic structures were formed became definitely established, they should be known as "crypto-explosion" structures.[6]

Although problems concerning meteorites and meteorite craters were receiving increasing attention, by no means everyone at mid-century was convinced that meteorite craters were to be found on the earth. To some competent geologists, the hypothesis was outrageous in the extreme and quite indigestible. N. H. Darton, for example, who had investigated various maars and other craters resulting from volcanic explosions, announced in 1945 that his geologic map of Arizona "was recently reprinted with repetition of the name Crater Mound for the wonderful rock upburst about 20 miles west of Winslow. . . . The words "Meteor Crater" are not admissible. I am convinced that no meteorite is present. . . ."[7] In 1946, nevertheless, the United States Board of Geographic Names officially changed the name from "Crater Mound" to "Meteor Crater." (The name "Barringer Crater" is also frequently used; in fact, the great hole was named the "Barringer Meteorite Crater" by the Meteoritical Society in honor of D. M. Barringer, who contributed so heavily in establishing its impact origin. And incidentally, Barringer was further honored in 1970, when his name was given to a crater on the far side of the moon.)

Darton was not alone in his opinion that the Arizona crater was the result of terrestrial forces. The matter was discussed at several professional meetings in 1949 and 1950, and a consulting geologist, Dorsey Hager, remarked in 1953 that a paper published by him could "justly be said to represent the opinions of numerous geologists and others interested in an objective explanation of the phenomenon."[8] To Hager, the Arizona crater did not appear to be the result of an explosion. He interpreted it as a graben-like sink, caused by solution of underlying evaporites, probably salt. (Hager had held this opinion for many years and had expressed it in 1926 in a letter to a mining journal—a letter which evoked a heavily sarcastic response from Baringer.[9]) He noted that the graben pattern together with sinkholes was a common feature in the surrounding region; and Crater Mound, as he continued to call it, was "the remnant or skeleton of an elliptical dome with a graben-like sink in its apex."[10] In his opinion the presence of meteoritic material was entirely coincidental—merely the result of a meteorite shower long after the formation of the crater.

These views brought heated responses from Nininger. Previously, in 1948, he had noted the reluctance of some geologists to admit the existence of terrestrial meteorite craters, and he remarked—without actually pointing out a parallel—that many years earlier the recognition by scientists of the extraterrestrial origin of meteorites themselves had been delayed by what he called a barrier of intellectual lethargy; and he listed a number of previously published facts which in his opinion clearly indicated that the Arizona crater had been formed by the impact and explosion of a meteorite. In 1954 he published a detailed criticism of Hager's paper, refuting point by point Hager's interpretation of the Arizona crater as an elevated hill rising above the plains with a deep hole in its top. Nininger had the support of Eliot Blackwelder of Stanford University, who took issue with Hager's suggestion that the silica glass (or lechatelierite) found at the crater had originated in explosive eruptions of volcanoes some 25 miles away, and also with his assertion that the "white sand" or "rock flour" found there in great quantities was derived by weathering from the sandy Kaibab limestone. On the whole, however, Hager's article caused fewer repercussions than might have been expected. By that time, in the minds of many investigators the origin of the Arizona crater seems no longer to have been a controversial matter; and a few years later, after seeing E. M. Shoemaker's maps of Meteor Crater, Hager reversed his opinion.[11] But there was still no easily applied test which could unerringly identify a geologic structure as having been formed by meteoritic impact. A new argument was needed, and in 1953 an apparently irrefutable one came from an unexpected direction.

In 1953, at the laboratory of the Norton Company in Worcester, Massachusetts, an industrial chemist named Loring Coes performed some experiments in an attempt to determine conditions necessary for the formation of certain naturally occurring minerals. He discovered that under specified conditions, one of which was extremely high pressure, a "new" substance was formed; it was later named coesite in his honor. He described it as a very dense, heavy form of silica.

Silica (silicon dioxide: $SiO_2$) occurs in nature as several distinct minerals. The differences between them are caused by differences in arrangement of the atoms of silicon and oxygen of which they are composed. The most widespread silica mineral is quartz, which is a constituent of many common rocks. Two other forms, or polymorphs, of silica are tridymite and cristobalite, both found in volcanic rocks. The rare silica glass, lechatelierite, in common with other glasses, has no highly ordered crystal structure. (In general, glasses form when the cooling of melted materials takes place too rapidly to permit the formation of crystals.) Opal, a form of silica which contains water, also lacks crystal structure. Coesite was found to be a dense, heavy crystalline material, a previously unknown form of silica.

Minerals found in the rocks of the earth's crust provide important clues as to conditions which prevailed at the time of their formation.

Geologists were therefore extremely interested in the discovery of coesite, and there was much speculation about whether it might occur naturally. Coes suggested that its existence in nature might have been overlooked because of its superficial resemblance to mica, a flaky mineral common in metamorphic rocks. Beause of its great stability, he had been unable to make any measurements of the reverse (coesite-to-quartz) transformation; and this suggested that rocks thought to have formed deep within the earth might contain coesite even after—due to one cause or another—they had reached the earth's surface. Samples of eclogite (a heavy rock originating at great depth) from South Africa, Austria, and Norway were carefully examined; and an eclogite inclusion found in a diamond mine in the Transvaal was studied.[12] No coesite was found. Coesite was also searched for, unsuccessfully, at sites of nuclear explosions.

In a book which appeared in 1956, Nininger suggested that a thorough search for coesite should be made at the Arizona crater:

> The Coconino sandstone which is found in the Arizona crater at depths of from 350 to 1000 feet is composed of quartz grains and is very porous. In view of the excessively high impact pressures involved in the formation of this crater, a thorough search for this [i.e., coesite] and possibly other new minerals both within the crater and under the southern rim, might have significant results.[13]

And before long, "new" minerals were discovered in the Arizona crater—minerals whose presence there was indeed highly significant.

Just a year earlier, in 1955, the laboratory creation of diamond by F. P. Bundy and his associates at the General Electric Company in New York sent a ripple of excitement through the scientific community. The formation of diamond was a venerable and obstinate mineralogical problem which for decades had fascinated scientists and non-scientists alike. It had long been known that diamond, the most brilliant of gems and the hardest known mineral, is composed entirely of carbon; and according to most authorities, it is formed under great pressure deep within the earth. Graphite, a black, opaque, relatively soft mineral, is also composed entirely of carbon. The difference between them is one of crystal structure; in other words, they are polymorphs of carbon. The tantalizing possiblity that graphite, if subjected to sufficient pressure, might be transformed into diamond, had led to much speculation and to a long series of interesting, but unsuccessful, experiments.

Bundy and his associates noted that earlier work on the thermodynamic stability of diamond and graphite indicated that at pressures between 400,000 and 1,400,000 psi (between 30,000 and 100,000 kilograms per square centimeter) and at temperatures in the range of 1000° to 3000°K, diamond is the stable form of carbon. During several years of experimental work, they succeeded in modifying an apparatus devised by P. W. Bridgman in such a way that these extremely high pressures and temperatures could be achieved simultaneously and maintained for periods of several hours. In this apparatus small but definitely proven diamonds were at last created.

This work recalled the diamonds observed by various people in the Canyon Diablo iron. "Could it be," wondered Nininger in 1956, "that the diamonds are actually formed on impact?"[14] Did their presence indicate that the enormous pressures and temperatures necessary for the laboratory creation of diamond had existed during the hypervelocity impact that formed the crater?

Nininger's speculations received little attention at the time, as a more popular theory was advanced by H. C. Urey in the same year. Urey made a valiant attempt to construct a history of the solar system which would account for the presence of diamonds in meteorites using, as he said, a completely new approach to the problem. He suggested that asteroids as large as the moon might have existed during the early history of the solar system, and that they were subjected to melting temperatures probably caused by radioactive heating. Gravitational pressures deep within such large bodies would have been great enough to cause formation of diamonds. And indeed, studies of the composition and structure of metallic meteorites by H. H. Uhlig indicated that processes through which they formed involved long, slow cooling from high temperatures under conditions of great pressure. Urey suggested that, after the final break-up of these moon-sized asteroids (probably due to collisions), fragments of them reached the earth as diamond-bearing meteorites. These ideas were well received by a number of widely recognized investigators. Controversy loomed, however, when detailed arguments supporting Nininger's view that the Canyon Diablo diamonds had been formed by impact were advanced by M. E. Lipschutz and E. Anders in 1960 (see Chapter 13).

An extensive structural analysis and remapping of the Arizona crater was undertaken during the 1950s by E. M. Shoemaker. He received the full cooperation of the Barringer Crater Company (formerly the Standard Iron Company) along with access to all parts of the great hole. He was even granted permission to remove rock samples. In May 1960, only a couple of months after Lipschutz and Anders stated their opinion that the diamonds in Canyon Diablo iron had been formed by the shock of impact, he sent one of these samples, a sheared fragment of Coconino sandstone, to E. C. T. Chao at the U.S. Geological Survey Laboratory in Washington, D.C.

Chao observed that the sheared quartz grains of the sandstone were embedded in a matrix of very fine-grained material. This material had been observed as early as 1907, when it was recorded by G. P.

Merrill, who thought it was opal, or some sort of silica glass. Chao investigated its index of refraction, a unique characteristic of every non-opaque mineral. He discovered that the index of refraction of the matrix material was higher than that of the fragmented quartz grains, indicating perhaps that the matrix was a very unusual glass, or possibly a mixture of extremely fine-grained materials. He made a standard X-ray powder diffraction pattern of the matrix material. The result indicated the presence of quartz—and of coesite.

To verify his discovery, Chao determined that the optical properties of the matrix material corresponded with those established for coesite. Further confirmation was obtained by comparison of his X-ray powder diffraction pattern of coesite with an X-ray powder diffraction pattern of coesite which had been synthesized by F. R. Boyd, of the Geophysical Laboratory of the Carnegie Institution in Washington. The two patterns were identical. In addition, a purified sample of the matrix material was subjected to a spectrographic analysis, which showed it to be composed of more than 99 percent silica. (Crystal structure is revealed by X-ray diffraction, making it possible to distinguish between polymorphs, as well as between different compounds, of the same elements. The elements present in a substance can be determined by spectrographic analysis.)

On the other side of the continent, in Menlo Park, California, Shoemaker received the news by telephone.[15] With a colleague, B. Madsen, he promptly subjected various specimens of Coconino sandstone, taken from different parts of the crater, to X-ray study. Coesite was found in compressed and sheared Coconino fragments taken from the mixed debris and breccia of the crater floor. It was found in the debris of the crater rim, and it was present in small amounts in the sandstone that had been converted to glass. It was also discovered in sandstone fragments taken from drill cuttings from depths of 600 to 650 feet beneath the crater floor. It was, in fact, identified as an abundant mineral[16] in the sheared Coconino sandstone of Meteor Crater.[17]

This was a startling discovery, with far-reaching implications. It also had some puzzling aspects. It clearly demonstrated that pressures and temperatures necessary for the transformation of quartz to coesite, pressures of at least 20 kilobars (20,000 atmospheres) and perhaps as much as 40 kilobars, along with temperatures in the range of 700° to 1700°C, had been generated by the shock of impact; but it was not immediately clear what other conditions were necessary for the formation of coesite, as certain apparently reasonable experimental procedures failed to produce it. Eventually it was concluded that meteoritic impact was probably the only natural event likely to occur on the earth that could provide the necessary pressure. This suggested that meteorite craters in quartz-bearing rock were locations in which coesite might be found—and that their impact

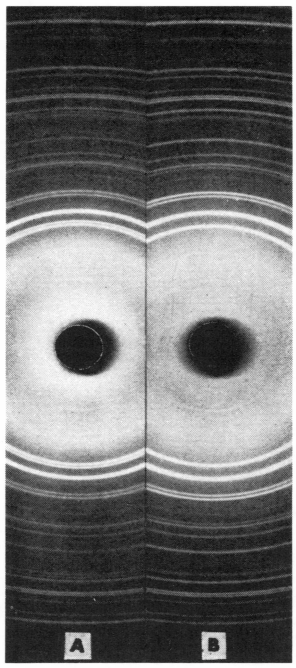

**10.1** *X-ray powder diffraction patterns of coesite.* A: *Natural coesite with minor amounts of quartz.* B: *Synthetic coesite.*

origin could be confirmed by its presence. Whether this would be true for very ancient meteorite craters would depend upon whether the stability of coesite proved to be such that it remained for long periods of time at ordinary atmospheric temperatures and pressures without reverting to quartz.

In the excitement of the discovery of naturally occurring coesite, various publications concerning suspected impact sites were reviewed, including several about the Ries Basin in Germany. Most investigators attributed it to some sort of volcanic action, possibly cryptovolcanism, or processes connected with the formation of maars. Shoemaker visited Europe in the summer of 1960. The trip was planned as a respite, following the death of his father, and he was accompanied by his mother and his wife. Nevertheless, it provided an opportunity to visit the Ries Basin, and there he was confronted by its puzzling features. Although much of the floor of the great structure is covered with recent lake sediments, he noted that breccias composed of old crystalline rocks were exposed in some of the hills. In some places, fragments of these rocks were mixed with fragments of Triassic shales and Jurassic limestone. In fact, unlikely mixtures of fragmented rock of various ages occurred in many places, and Shoemaker noted that some of the very early investigators had concluded that an explanation of the manner in which they had been emplaced would solve the problem of the origin of the Ries.

A remarkable feature of the Ries Basin was the presence of suevite, a peculiar rock which was plentiful in some parts of the structure but had never been found at any other location. It was somewhat similar to tuff, a pumice-like rock composed of very small, compacted, and often glassy volcanic fragments. Suevite had been interpreted as a volcanic rock, and the belief held by many investigators that the Ries Basin was volcanic was based upon its presence. Shoemaker, however, found nothing to indicate that it was volcanic. On the contrary, he found that it consisted of shattered and sintered fragments of many different kinds of rock embedded in various kinds of glass. In his opinion it was a compact and glassy breccia, and it bore remarkable similarities to the coesite-bearing Coconino sandstone of Meteor Crater. After visiting many quarries and outcrops of suevite, he mailed a sample of it to the U.S. Geological Survey in Washington, where it eventually reached E. C. T. Chao. Chao examined it and discovered that it contained coesite.[18]

Shoemaker observed that the meteoritic origin of the Ries Basin, indicated by the presence of coesite, was confirmed by various other features. Structurally it resembled no known volcanic caldera or crater. Even in the most violent of known volcanic eruptions there had been no ejection of rock masses as large as the great overlapping thrust slices of the imbricated zone (known as the Schollen-und-Schuppen

**10.2   A section of the rim of the Ries Basin, showing a quarry.**

Zone) where huge slices of rock partially covered one another like the scales of an enormous fish. In Shoemaker's opinion, the major structural features which had puzzled so many investigators appeared to have a straight-forward explanation in terms of hypervelocity impact mechanics.

A third occurrence of coesite was recorded in 1961. It was found in two samples of material collected at Wabar by Virgil Barnes, of the University of Texas. One sample consisted of white siliceous material enclosed in black glass. The other was a piece of fractured sandstone. Investigators at the U.S. Geological Survey Laboratory reported that X-ray patterns of coesite from Wabar "agree in every detail with those of synthetic coesite and natural coesite from Meteor Crater, Arizona."[19]

In 1933 H. P. T. Rohleder was one of very few supporters of the idea that the Steinheim Basin in Germany was the result of meteoritic impact. Nevertheless, he agreed with most other authorities in assigning a volcanic origin to the nearby and much larger Ries Basin, because of the presence in the Ries of suevite. In 1936 he came to the conclusion that the crater occupied by Lake Bosumtwi in Ghana (Ashanti) was created by volcanic action and not, as claimed by Malcolm Maclaren a few years earlier, by meteoritic impact. Rohleder's opinion was based on his discovery of a peculiar rock, very similar to the suevite of the Ries Basin, in the Bosumtwi crater. He remarked that "the key to the correct interpretation of the Bosumtwi depression is to be found in the Ries of Nördlingen,"[20] and he noted a number of similarities between the two structures. In the light of later work, these similarities assume added significance—for in 1961 the suevite-like rock from the Lake Bosumtwi crater was shown to contain both lechatelierite and coesite. This definitely established the Bosumtwi depression as a meteorite crater, and as the fourth location in which coesite had been found.[21]

During the years following the discovery of natural coesite, its presence became widely accepted as an indicator of meteoritic impact. Before the end of 1961, coesite was reported in two structures previously thought to be cryptovolcanic, the Kentland quarry in Indiana and Serpent Mound in Ohio.[22] Questions as to its stability were apparently resolved when the discovery of coesite was reported in fragmented Precambrian rocks taken from the Holleford crater in

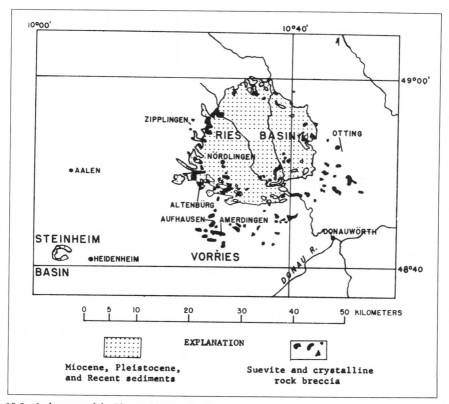

*10.3   Index map of the Ries Basin, Bavaria, Germany.*

**10.4** *Shatter cone from the Kentland quarries, Indiana. Flaring striations distinguish shatter cones from slickensides, which display parallel grooves.*

Ontario, Canada, where a shallow circular depression was first perceived in aerial photographs. Investigation revealed that a bowl-shaped depression, estimated to have been formed between five and six hundred million years ago, existed in the underlying Precambrian rock. This, it was pointed out, was the "oldest occurrence of coesite known at this time, and it established the fact that coesite may persist, at least in part, in a moderate temperature-pressure environment, through geologic time. . . ."[23]

And as if this were not enough for one year, still another discovery was made in 1961. A new dense polymorph of silica, which required for its formation even greater pressures and temperatures than those needed for the formation of coesite, was synthesized by two Russian investigators, S. M. Stishov and S. V. Popova. The "new" material was named stishovite; but as it turned out, this honor unfortunately did not please the Russians, as had been intended, for Stishov was the junior author![24] It was identified in coesite-bearing Coconino sandstone from Meteor Crater the following year[25]; and its presence gave further support to the idea that the diamonds in Canyon Diablo iron had been formed by the shock of impact.

It is interesting, incidentally, to note that the peculiarities of the shocked Coconino sandstone had been observed by D. M. Barringer in 1909:

> what we have termed Variety A consisted of angular fragments
> of undoubtedly original sandstone still showing the bedding

planes, but often with strongly marked slaty structure at more or less right angles to the bedding planes which . . . can be due only to the exercise of very great lateral or perhaps radial pressure. . . . Dr. J. W. Mallet was the first to assure me that this rock was certainly of terrestrial origin . . . since I had never seen a rock like it. . . .[26]

While this work was in progress, another possible means of identifying meteorite craters was being investigated by R. S. Dietz. His interest in impact structures dated from his student days at the University of Illinois, when he made frequent trips to the Kentland quarry. The shatter cones to be found there ranged in length from less than an inch to more than six feet. His observations of their orientation resulted, as we have seen, in his opinion that they were formed by forces which came from above rather than from below[27]; and that, as suggested by Boon and Albritton, structures similar to the Kentland disturbance—the so-called cryptovolcanic formations—were the "roots" of ancient meteorite craters.

Dietz defined shatter cones as "striated percussion fracture cones" or, more simply, as "conical fragments of rock characterized by striations that radiate from the apex." In his opinion, cryptovolcanic or crypto-explosion structures, which he called "astroblemes"—literally, "star wounds"—were created by the hypervelocity impact of large meteorites. Hypervelocities he defined as velocities greater than the speed with which sound travels through average rock; greater, that is, than about 5 kilometers (3 miles) per second. It is thought that, on the average, a giant meteorite strikes the earth with a velocity of at least 15 kilometers (9 miles) per second. Hypervelocity impact must create an intense, high-velocity shock (or pressure) wave which spreads out from the impact point through the surrounding rocks. Therefore, Dietz pointed out, indications that a large volume of rock has been intensely and naturally shocked would constitute evidence of meteoritic impact. Perhaps the presence of shatter cones could provide such evidence.

It was not known whether shatter cones could be formed by any force other than a shock wave. They had never been discovered in "normal" geological formations, nor in rocks known to have been subjected to volcanic explosions. Explosives such as dynamite, used in quarrying operations, produced rude and irregular fracture cones which lacked striations and displayed larger apical angles than those of shatter cones. Military explosives of high detonation velocity and high shattering effect produced cones, less perfect than shatter cones in shape, but with similar striations. In 1959 shatter cones were discovered in volcanic tuff at the site of an underground nuclear bomb test, and were attributed to the shock created by the nuclear explosion. But shatter cones had been reported from only a few of the known cryptovolcanic structures. In fact, as of 1959, they were known only in the following locations: the Steinheim Basin in Germany, where they were first observed; the Bosumtwi crater, as reported by H. P. T. Rohleder; and in three locations in the United States—the Kentland disturbance in Indiana, Wells Creek basin in Tennessee,

10.5   *Dolomite from Wells Creek basin, with shatter cones showing uniform orientation. The cones are interlaced, and new fractures of the rock reveal new shatter cones (About ½ actual size).*

10.6   (Opposite) *Oblique aerial view of the Steinheim Basin in southern Germany. Note central uplift in center of crater with the town of Steinheim on the left side. The eroded crater walls surround farmed areas of rectangular fields on the crater floor. View is looking northeast.*

and the Crooked Creek structure in Missouri (to which C. W. Wilson in 1953 and H. E. Hendriks in 1954 ascribed a meteoritic origin).

In 1946, during a visit to the Wells Creek basin to observe the orientation of shatter cones, Dietz found no good bedrock exposures in which they occurred. The great displaced blocks of rock that did contain shatter cones were so jumbled and disturbed that it was impossible to determine their original position. Similarly, in the Steinheim Basin, which he visited in 1956 and again in 1957, bedrock exposures were poorly developed, and in addition were usually inaccessible, as they were found mainly in excavations made for houses. Great limestone blocks had been displaced in the disturbance which created the basin, and because of this it was impossible to discover whether the shatter cones they contained were oriented in any common direction. Dietz also visited the Ries Basin, only twenty-six miles from Steinheim. He discovered no shatter cones there but was struck by the similarity of the two structures. He suggested that they had been formed at the same time in a double holocaust.

In October 1959 several scientists, including R. S. Dietz and E. M. Shoemaker, were invited by G. Kuiper to visit the MacDonald Observatory at the University of Texas to study the surface of the moon. An interruption in this study caused by cloudy weather provided an opportunity to visit the nearby Sierra Madera dome. This

10.7   *Shatter cone from the Sierra Madera, Texas (Actual size).*

formation, described by P. B. King in 1930, was included by Boon and Albritton in their list of structures probably created by meteoritic impact. Dietz reported the frequent occurrence of shatter cones at the Sierra Madera site; in fact, the shatter coning proved to be more extensive than at any other location where shatter cones had been found.

This discovery stimulated further search. November 1959 found Dietz and Shoemaker at Serpent Mound, the highly deranged structure in Ohio first mapped by Bucher in 1920. When a quarry containing great masses of broken rock revealed no shatter cones, an intensive search was made in the precise center of the structure. As there were no exposed outcrops, they proceeded to break open what boulders they could find. In a period of two days, shatter cones were discovered in a number of boulders, occurring in approximately one of every twenty-five of the boulders that were split.

Flynn Creek, Tennessee, was visited next. This structure too was included in Boon and Albritton's list of probable impact sites. Late in November 1959 Dietz, accompanied by C. W. Wilson and R. Stearns, discovered a thin bed of poor but unmistakable shatter cones in a new road cut near the center of the Flynn Creek structure. This discovery brought the total number of shatter-cone locations then known to eight.

No shatter cones were found at Jeptha Knob or at Howell, Tennessee. And surprisingly, at Meteor Crater, Arizona, a search for shatter cones went on for several years before a small shatter-coned fragment of Coconino sandstone was discovered in the debris of the crater rim by E. C. T. Chao. It was a poor sample, however, and doubts have been expressed, even by Chao himself, as to its authenticity. The doubts were shared by Shoemaker and others who described it as "a piece of Coconino sandstone that exhibits part of a curved slickensided surface rudely resembling part of a shatter cone."[28] As a possible explanation of the absence of shatter cones in what many thought to be the most likely site of all, it was noted that the rocks immediately beneath the explosion center—"ground zero," the most probable location for shatter cones—were now covered with lake beds. Eventually, perhaps, shatter cones might be discovered below the lake sediments.

Thus, by the early 1960s, two distinguishing features by which evidence of meteoritic impact might be determined were accepted, in some cases tentatively, by many investigators. But problems remained. Shatter cones were not well understood. Although they had been found only at sites of suspected meteoritic impact, proof that they could not be formed in other circumstances was lacking. In fact, in 1961, well-formed shatter cones were created when aluminum pellets were fired at high velocity into Kaibab dolomite.[29] And although the early search for coesite in deep-seated rocks had been unsuccessful, in 1961 it was discovered as an inclusion in diamonds.[30] It began to appear that reliable criteria for the identification of meteorite craters might be reached by making use of the suggestion of Dietz and others that evidence of impact could be safely drawn from large volumes of rock which had been naturally and intensely shocked.

## CHAPTER 11

# Meteorite Craters in Canada and Crater Forms

WHILE INVESTIGATIONS CONCERNING impact phenomena were going on in the United States during the 1950s, a search for ancient impact sites was undertaken in Canada as a result of the great interest aroused by the discovery of the New Quebec crater. The search was highly successful. A number of "fossil" meteorite craters were discovered in a remarkably short time, and study of them contributed greatly to development of criteria by which meteorite craters may be recognized.

In 1943 a small and remarkably round lake near Hebron, Labrador, was photographed by Colonel A. F. Merewether of the U.S. Army Air Force. It attracted no particular attention at the time. Later, however, word of its existence reached V. B. Meen as a consequence of his work in connection with the New Quebec crater. In 1953 Meen made a search by air in the region of the Labrador-Quebec boundary at latitude about 58°N, and eventually located the crater.[1] Although bad weather prevented detailed exploration, Meen suspected it to be of meteoritic origin. In 1954 a second trip was made, and a study of the crater and its surroundings was undertaken.

The rock of the region was Precambrian with a covering of coarse glacial till. Neither volcanic rocks nor meteoritic fragments were discovered. Meen admitted that he found no irrefutable evidences of meteoritic origin; but among the evidence which did exist were the remarkable symmetry of the crater, its raised edges, and its excellent agreement with Baldwin's law for explosion craters. Possible origin by the melting of a pocket of glacial ice was discounted, as that would not explain its circular shape; and "it would seem," he remarked, "that an origin by solution of underlying bedrock is most unlikely and one by

*11.1    Brent impact structure, Ontario, Canada. Diameter of the visible circle is almost 10,000
feet (3,050 m).*

volcanism impossible."[2] The crater became known as the Merewether,
or Hebron, crater. Two very similar craters were found nearby,
suggesting that all three might have been caused by a shower of
meteorites.

A discovery which greatly influenced later work on meteorite
craters was that of the Brent crater in Ontario. It was first noted in
1951, by J. A. Roberts, president of the Spartan Air Services Ltd. of
Ottawa. His company had arranged to take high altitude photographs
of parts of Ontario for the Canadian government; and in examining
aerial photographs of a location near Brent, close to the northern edge
of Algonquin Provincial Park, Roberts observed what appeared to be
an almost perfect circle outlined in a region of thick forest and small
lakes. The photograph was shown to C. S. Beals, the Dominion Astronomer, with the result that in the summer of 1951 an expedition
was sent to study the area.

Although it was inaccessible by road, the small village of Brent
could be reached by railroad; from there, old logging trails led to
Gilmour Lake, one side of which followed part of the circular outline.
The circle observed in the aerial photographs had a diameter of almost

10,000 feet (3,050 m). No abrupt crater-like depression was discovered, though the central part of the circle was lower than nearby hills. It was remarked that "to an observer on the ground there is no indication that the feature exists. In all probability it would have been discovered without the assistance of high altitude air photography."[3]

The circular structure lay in a region of ancient granite and gneiss. Although explorations were hampered by thick vegetation, several undisturbed outcrops of bedrock were found at nearby points outside the circle. None was found inside. This fact, it was thought, perhaps indicated that the circle was the outline of an ancient crater now filled with later rock. In several places near its circumference, outcrops of fragmented rock were found which, according to P. M. Millman, "had somewhat the appearance of consolidated explosion breccia."[4] Chunks of cemented breccia were also found within the circle, and some of them contained blocks of gneiss several feet across. In addition, large blocks of limestone and shale of Ordovician age were found along ridges in the center of the basin and also along the margin of one of the lakes. If, as seemed to be increasingly indicated, the formation was the eroded remnant of an ancient meteorite crater, the rim of rock fragments which must have originally surrounded it had no doubt been worn down by the leveling action of glaciers.

The investigators were well aware that circular features on the surface of the earth may result from any of a number of causes. They may be produced by various volcanic processes, by erosion basins, by salt domes, by frost action, by sinkholes—to name a few possibilities. If indeed the Brent crater was created by the impact and explosion of a meteorite, its original form, as Millman remarked, had "been greatly modified, and in fact almost obliterated by subsequent changes on the earth's surface."[5] However, the evidence suggested that a meteoritic origin was probable, and an intensive study was begun.

Investigations by the Geological Survey of Canada and the Dominion Observatory continued for several years. A contour map of the area was constructed on which the circular feature stood out very clearly, and it was observed that the circle appeared to be quite unaffected by structural trends in the surrounding gneiss.

Various geophysical techniques were used to determine the substructure of the formation. Gravity investigations were made, as small variations in gravity measurements can indicate differences in the density of subsurface materials; the forces of gravity were found to diminish slightly in a concentric pattern toward the approximate center of the circle. A magnetic study revealed that magnetic intensities also diminished in a similar concentric pattern toward the center. Besides gravity and magnetic studies, a third type of investigation was undertaken employing seismic methods.[6] The results of these investigations confirmed suspicions that the Brent feature was the outline at ground surface of a deep, steep-sided depression in the regional rock, filled with rock of a different type, and with a slightly heavier layer at the bottom. Subsurface exploration by diamond drilling was undertaken,[7] the first large-scale diamond drilling undertaken in Canada for purely scientific purposes.[8] It revealed that below a fairly thick cover of glacial debris the depression was filled with sedimentary rocks, chiefly limestone and sandstone, which rested on a bottom layer of breccia composed of fragments of the heavier gneiss. Below the breccia lens was a zone of fractured rock. Some of the drill cores contained bits of breccia so small and breakable that they could be crumbled in the fingers. In others, particularly those taken from the upper part of the breccia zone, the fragments of gneiss were strongly bound together with a glassy cement, which was interpreted as a melt indicating that the material had been subjected to extreme heat. But although the breccia provided a feature in common with known meteorite craters, it did not provide proof of meteoritic origin.

The sedimentary rock which filled the basin contained marine fossils of Ordovician age, indicating that the rock-forming sediments had been deposited in Paleozoic seas that covered the region more than 400 million years ago. The ancient gneiss of the surrounding area is thought to have an age of not less than 800 million years. The depression, or crater, must therefore have been created during the immense timespan between about 400 and at least 800 million years ago. A probable age of about 500 million years seems to be generally accepted.

Following these discoveries, interest in terrestrial craters was further stimulated by the continuing controversy about the craters on the moon. An eloquent exponent of theories which attributed the lunar markings to processes originating within the moon itself was J. E. Spurr. There was a gradually growing opinion, however, that no one theory was likely to explain their entire range and complexity. That some small volcanic formations existed on the moon seemed indisputable, but an increasing number of observers began to share the opinion that most lunar craters were meteoritic. Arguments in support of this view were advanced by various investigators, in particular, as already mentioned, by R. S. Dietz in 1946 and R. B. Baldwin in 1949. Strong additional support came from G. P. Kuiper in 1954 and from H. C. Urey in 1956.

It was generally agreed that if the moon had indeed undergone heavy meteoritic bombardment in the geologic past, the earth could not have escaped similarly numerous impacts. Perhaps the New Quebec and Brent craters were evidence of such events. Although these recently discovered features were not immediately proved to be meteoritic, they exhibited significant resemblances to known meteorite craters. Their bowl-like shape with steep inner sides, the presence of great quantities of fragmented rock indicative of an explosion, the complete absence of volcanic material—all this, along with the lack of any other acceptable hypothesis of origin, convinced many (though by no means all) investigators that these craters must be the result of meteoritic impact. The fact that both of them lay on the Canadian Shield—that enormous expanse of Precambrian rock covering about half the area of Canada—was noted with interest by a number of Canadian scientists.

The Canadian Shield is estimated to cover some 1,800,000 square miles, and most of it is thought to have been geologically undisturbed for hundreds of millions of years. If the earth was once subjected to a meteoritic bombardment similar to that which pock-marked the moon, some indications of it might still, perhaps, remain in this ancient rock. The fact that the existence of both the New Quebec and Brent craters was first perceived on aerial photographs suggested that a search of aerial photographs taken of the vast region of the Canadian Shield might result in discovery of other similar craters.

At that time, the Canadian Air Photo Library contained some two and a half million photographs, covering most of the mainland area of Canada. A systematic search through them began in June 1955. The photographs were carefully scrutinized, and occasional circular formations were examined with a stereoscope to detect any resemblance to lunar craters or to terrestrial meteorite craters. In general this meant a search for circular depressions with raised rims, similar to the new Quebec crater or the Arizona Meteor Crater. However, as noted by C. S. Beals, M. J. S. Innes, and J. A. Rottenberg,

> the search had proceeded for a considerable time and had involved the examination of several hundred thousand photographs before it was finally realized that such conspicuous features were practically non-existent and that a really effective search would have to concentrate on less obvious aspects of crater structure. This conclusion was reinforced by the discovery of three circular features in which raised rims were inconspicuous, though not entirely lacking, and which were recognized as explosion craters by their underground features when studied by geophysical methods and diamond drilling techniques.[9]

The three circular features referred to were the Brent and Holleford craters in Ontario, and the Deep Bay crater in Saskatchewan. The Brent crater had no conspicuous rim, though some indications of one were discerned, considerably worn down and in some places almost obliterated by glaciation. The Holleford crater was first perceived during examination of some 200,000 aerial photographs, mainly of regions in Ontario and Quebec. It was considered the "most promising" of several circular formations discovered soon after the search began. In most places in the vicinity of Holleford, however, the Precambrian gneiss lay beneath ancient sedimentary rock of marine origin. Sedimentary rock overlay the circular formation, where the strata were found to dip radially inward. A somewhat subdued saucer-shaped indentation could be perceived from the air. Geophysical investigations similar to those made at the Brent crater indicated that a deep bowl-shaped cavity in the Precambrian basement rock lay beneath what C. S. Beals described as a moderately inconspicuous circular depression at the ground surface. Vestiges of the original rim were eventually located by drilling, at a depth of sixty-four feet. In addition, drilling revealed that several hundred feet of shattered and pulverized rock lay beneath the floor of the depression. (It was from drill cores taken from this crater that the oldest occurrence of coesite known at that time was reported in 1961.[10]) Beals and his associates suggested a probable age for the crater of 500 to 1,000 million years.

The Deep Bay crater in northern Saskatchewan was observed from the air in 1947 by M. J. S. Innes, who described it as a large, water-filled depression at the southeast end of Reindeer Lake. Other observers, too, had commented on its unusual features. In an account of a canoe trip in northern Saskatchewan, P. G. Downes remarked in 1943 that

11.2   Holleford impact structure, Ontario, Canada.

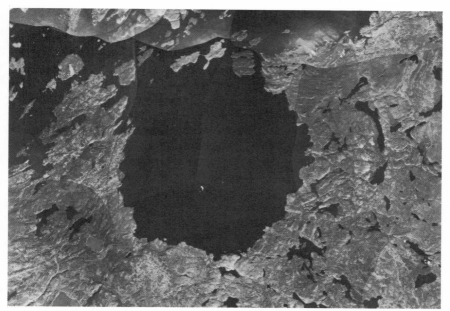

11.3    Vertical aerial photomosaic of Deep Bay crater, Saskatchewan, Canada. Diameter of
        the lake is about seven miles (about twelve kilometers). Abrupt changes in shading are
        mosaic edges.

in the southeast corner [of Reindeer Lake] is a large and perfectly symmetrical round bay quite devoid of islands and tremendously deep. This huge bay, utterly out of character with the rest of the lake, is a spot usually to be avoided. It is too deep for nets, affords no shelter from the wind, and is supposed to be inhabited by a gigantic fish of miraculous abilities and voracious appetite, inclined occasionally to come up through the ice and select itself a young caribou.[11]

The first suggestion that Deep Bay might have been formed by a meteorite was made in 1951 by D. S. Rawson, of the University of Saskatchewan.[12] Rawson was aware of the recent interest in craters of possibly meteoritic origin and noted the remarkable circularity of the bay. He recorded a depth of 660 feet there, about three times the depth of any other part of Reindeer Lake. In his opinion the bay was a crater of some kind. Rawson's observations unfortunately did not come to the attention of the Dominion Observatory. However, as interest was being increasingly stimulated by the investigation of other crater-like features, attention eventually turned to Deep Bay, and in 1956 an expedition was organized to examine it. Topographical, geophysical, and geological studies were undertaken, and the results indicated that the feature was caused by a tremendous explosion which probably occurred sometime before the last period of glaciation. No rim remained, and no meteoritic material was found. There was no indication of volcanic action. Much shattering and fragmentation of the rock could be seen on the inner walls, and surface rock around the edge of the crater was deeply fractured, a condition which decreased in intensity with increasing distance from the edge. A meteoritic origin for the crater was suggested.

In 1956 a report on the progress of the search for "fossil meteorite craters," or, as they put it at the time, for analogies between lunar and terrestrial topography, was published by C. S. Beals, G. M. Ferguson, and A. Landau. Their list included several circular features which had been brought to their attention either by other interested observers or through the search of aerial photographs. A number of them had not been studied and were included as possible locations for future investigation. Several of the circular features were thought to be volcanic; but it seemed probable that the Clearwater Lakes and Macamic Lake in Quebec, and a crater near Franktown in Ontario, in addition to the New Quebec, Brent, and Holleford craters, were meteoritic. A meteoritic hypothesis was soon discounted, however, for both Macamic Lake and the Franktown crater, in the former case because of results of a gravity survey, and in the latter because of evidence obtained by drilling. But the Clearwater Lakes—two roughly circular bodies of water separated by a screen of islands—stood out conspicuously in a region where most of the numerous lakes were elongated in the same direction, apparently as a result of glaciation. The circularity of the two lakes suggested an impact origin, and further investigation was recommended. But the writers pointed out that "the finding of such circular features on aerial photographs is only the first step in a lengthy and uncertain process. For any but geologically recent objects it is unlikely that the more obvious criteria of crater recognition can be applied. . . ."[13] They noted, in addition, that the expenses involved in geophysical studies and drilling operations were enormous, and they suggested that a few typical craters might be extensively studied and then used as standards of comparison.

Largely because of the search of aerial photographs, some half dozen suspected meteorite craters had been discovered in Canada in less than a decade. Although these various craters exhibited indications of meteoritic origin which convinced many investigators, proof was still lacking. The acceptance of meteoritic origin for any crater in the absence of proof had already drawn severe criticism, as in the following example:

**11.4**   *Vertical photomosaic of Clearwater Lakes, Quebec, Canada. Diameter of the larger lake is about 20 miles (33 km).*

Recently a considerable number of crater-like topographic features have been asserted, more or less positively, to be due to meteoritic impact. Critical examination of the bases for such identification in the case of the most widely publicized features of this sort discloses no valid reason for accepting the meteoritic nature of the craters in question. The most flagrant case is that of Chubb Crater in Canada, where (in the complete absence of meteorites and also of materials definitely identifiable as produced by the heat and pressure incident to meteoritic impact) the asserted occurrence of a magnetic anomaly, in rock admittedly rich in segregations of magnetite and allied minerals, is advanced as "proof" that the feature is a meteorite crater.[14]

The writers went on to assert that the authentication of a meteorite crater should rest either on discovery in or near the crater of meteoritic material or altered material known to have resulted from meteoritic impact (such as lechatelierite), or on actual observation of meteorite falls having impact points in the crater area.

Some of the craters discovered on the Canadian Shield were of immense age—hundreds of millions of years old. They had been subjected to glaciation and other kinds of erosion during vast geological ages; and as Boon and Albritton remarked some twenty years earlier, meteoritic materials would be the first to disappear under the on-slaught of erosion. It therefore seemed necessary to determine other criteria by which a meteorite crater could be identified. This problem was attacked by Beals, Innes, and Rottenberg, who recognized that the first requirement in any search for ancient meteorite craters is "a knowledge of the complete structures of explosion craters . . . not only those parts of the craters visible on or above the earth's surface, but also the subsurface parts of the structure, since for many objects this subsurface portion may be all that is left to observe."[15]

During the late 1950s there was considerable investigation of craters resulting from nuclear explosions—the only man-made ex-plosions releasing forces comparable to those involved in meteoritic impact. After study of the available literature, Beals and his associates reached the conclusion that in the case of a meteoritic impact in granitic gneiss, some vaporization and melting of the rock would occur at the center of the explosion, and a mass of fragmented rock would remain at the bottom of the crater to a depth approximately one-third of the crater diameter. Below that, a region of fractured rock would extend downward for at least another one-third of the crater diameter, and for possibly twice that depth in the direction in which the meteor-ite struck.

On the basis of what they determined to be a typical meteorite crater they suggested a number of possible forms in which visible remnants of such a crater might remain on the surface of the earth after the passage of millions of years. Even though the rim of a meteorite crater might have been removed by erosion, the crater itself might still form a conspicuous feature on the landscape, such as a circular lake. Deep Bay, Saskatchewan, might be an example of this type of remnant. A crater which had never been covered with water might be almost obliterated by ordinary processes of erosion, but might nevertheless be revealed on aerial photographs by drainage patterns or by configurations of vegetation. Or, beneath a thin cover of sedimentary rock, the shape of a buried crater might still be de-tected, as at Holleford. Sediments covering a crater might at some later time be removed, revealing a circular outline, as at the Brent crater. Another possibility might be a circular arrangement of up-turned strata in sedimentary rocks. Still another might be the com-plete destruction by erosion of the upper parts of a crater, leaving only a circular deposit of breccia to give a clue to its past existence. And finally, a meteoritic impact of sufficient violence might trigger the release of volcanic forces, which would complicate interpretation of the event.

In response to statements that the presence of nickel-iron was the only authoritative criterion for the identification of a meteorite cra-ter, the investigators countered with the remark that it may well be that a more fundamental indication of a crater's meteoritic origin is the underground structure of crushed and shattered rock associated with it. On the basis of modern theoretical ideas . . . it appears more and more unlikely that any other natural process can produce the distribution of melted, crushed, and shattered rock that is characteristic of a large meteorite crater.[16]

Beals and his associates also noted that the possible existence of eroded remains of ancient meteorite craters on the surface of the earth was first suggested in connection with cryptovolcanic structures. The origin of these structures was still being debated: were they the eroded scars of meteoritic impact or of ancient subsurface volcanic processes? In the opinion of Beals *et al.*, the consistent presence of large quantities of shattered and pulverized rock at the various cryptovolcanic sites, together with the absence of any evidence of volcanism, strongly suggested meteoritic impact. In support of this argument, they noted that the impact of a meteorite upon the surface of the earth involves pressures many times greater than those involved in volcanic explosions. They referred to the work of K. E. Bullen, according to whom "the strengths of the Earth's materials set an upper limit to the distortional stress, but not to the pressure that can be present in the interior."[17] In other words, beyond certain limits, the rocks of the earth's crust would be unable to contain volcanic activity in which very high pressures build up. Long before volcanic pressures reached peaks similar to those involved in the impact of a meteorite which had been traveling at cosmic speed, they would be released by volcanic action such as a lava flow or a volcanic explosion. And in volcanic explosions, as Beals *et al.* pointed out,

> intense brecciation is pretty well confined to the area of the vent where the explosive forces burst through the Earth's surface and does not, as in the case of meteoritic impact, extend into the country rock to distances of the order of 10−20 times the diameter of the exploding body. It seems probable therefore that detailed gravity observations combined with diamond-drilling techniques should eventually make possible a definite decision as to the origin of cryptovolcanic structures.[18]

In neither of the two papers published in 1960 and 1963 by Beals, Innes, and Rottenberg is there any mention of coesite. No doubt this omission occurred because at that time events were taking place in such rapid succession that publication inevitably lagged behind. Their paper entitled "The Search for Fossil Meteorite Craters" appeared in June 1960 in *Current Science*, published in Bangalore, India. Discovery of the first natural occurrence of coesite was also

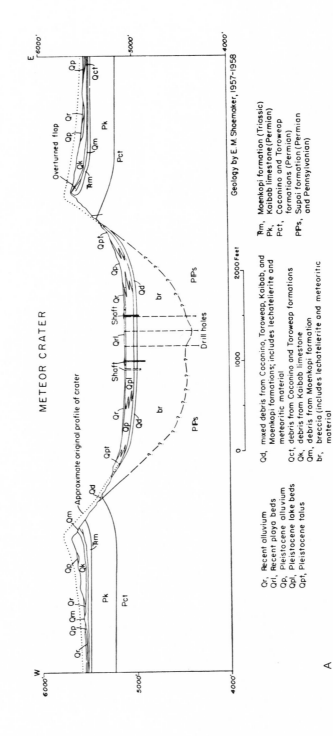

# METEOR CRATER

Approximate original profile of crater

Overturned flap

Shaft

Drill holes

Geology by E. M. Shoemaker, 1957–1958

W
6000'
5000'
4000'

E
6000'
5000'
4000'

0    1000    2000 Feet

Qr, Recent alluvium
Qrl, Recent playa beds
Qp, Pleistocene alluvium
Qpl, Pleistocene lake beds
Qpt, Pleistocene talus

Qd, mixed debris from Coconino, Toroweap, Kaibab, and
    Moenkopi formations; includes lechatelierite and
    meteoritic material
Qct, debris from Coconino and Toroweap formations
Qk, debris from Kaibab limestone
Qm, debris from Moenkopi formation
br, breccia (includes lechatelierite and meteoritic
    material

Ŧm, Moenkopi formation (Triassic)
Pk, Kaibab limestone (Permian)
Pct, Coconino and Toroweap
    formations (Permian)
PPs, Supai formation (Permian
    and Pennsylvanian)

A

11.5 *Comparison of crater rims: A: Meteor Crater. B: Nuclear explosion craters, Teapot Ess and Jangle U.*

NUCLEAR EXPLOSION CRATERS

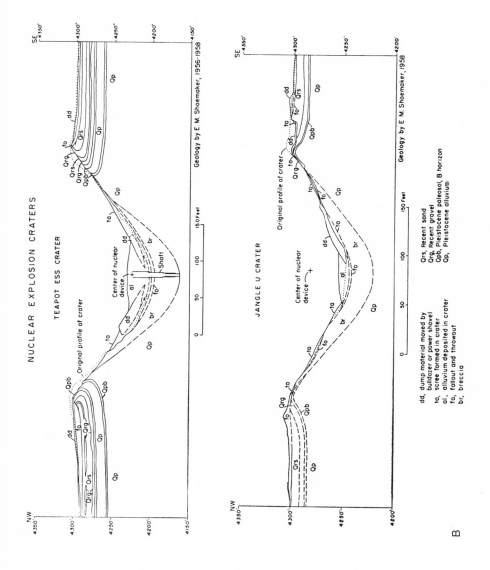

TEAPOT ESS CRATER

Geology by E. M. Shoemaker, 1956-1958

JANGLE U CRATER

Geology by E. M. Shoemaker, 1958

dd, dump material moved by
     bulldozer or power shovel
ta, scree formed in crater
al, alluvium deposited in crater
fo, fallout and throwout
br, breccia

Qrs, Recent sand
Qrg, Recent gravel
Qpb, Pleistocene paleosol, B horizon
Qp, Pleistocene alluvium

B

announced in June 1960. By then, however, their second paper, "Fossil Meteorite Craters," a slightly expanded version of the first one, was apparently already in press. It was included in the fourth volume of a series on the solar system entitled *The Moon, Meteorites, and Comets,* edited by B. M. Middlehurst and G. P. Kuiper; but because of the immense labor and unavoidable delays involved in assembling a large and comprehensive work, the fourth volume did not appear until 1963.[19] In the meantime in 1961, as we have seen, coesite was discovered in suevite from the Ries Basin, in glass and rock fragments from the Wabar crater, in suevite-like rock from the crater of Lake Bosumtwi, and in two cryptovolcanic structures in the United States (the Kentland quarry and Serpent Mound) —providing proof, in the opinion of many investigators, that these locations were scars of meteoritic impact. Most significant of all, with respect to the ancient craters of the Canadian Shield, coesite was reported in drill cores from the Holleford crater in Ontario. A brief announcement of this discovery was made in 1961, too late for inclusion in the Middlehurst and Kuiper volume.[20] It was not until 1963 that a more widely circulated account of it appeared in *Science.*

While the search for fossil meteorite craters was going on in Canada during the 1950s, an investigation of crater forms was being carried out in the United States, stimulated by the possibility of some day placing human beings on the moon. This work involved a search for similarities between terrestrial and lunar craters, and a comparative study of terrestrial maars and meteorite craters—the two crater types occurring on the earth which were considered to be most similar to lunar forms, and which sometimes appeared so similar to each other as to be difficult to distinguish. Much of this investigation was carried out by E. M. Shoemaker, who remarked in 1962 that

> there is nothing new in the thesis that close comparison of lunar craters with possible terrestrial homologues will prove fruitful, for it has been followed in lunar studies since the time of Galileo. The justification for taking it up anew is that our understanding of terrestrial phenomena relevant to the moon has advanced greatly over the course of the years, particularly in the last two decades.[21]

Shoemaker was well acquainted with regions in the southwestern United States where maars occur. In 1956 he described the geologic setting of uranium deposits which he had discovered in diatremes on the Navajo and Hopi reservations in Arizona, New Mexico, and Utah. This work involved careful study of the diatremes, or "explosion pipes," which, as we have seen, culminate at ground surface in maar-type volcanic craters. At least 250 maars are well exposed in the Hopi

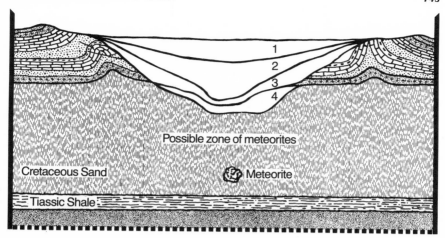

Possible zone of meteorites

Cretaceous Sand

Meteorite

Tiassic Shale

11.6 Section across the largest of the Odessa craters (crater No. 1). Its diameter is 550 feet (150 m).

Buttes region, more than are known in any other area of comparable size. Having been subjected to erosion for several million years, many of the diatremes were found to be stripped down, dissected, or even partially destroyed, and thus presented an excellent opportunity to observe the substructure of maars. In addition to this work, Shoemaker took part in a program of crater investigation which involved a structural analysis and remapping of Meteor Crater, and in 1959 he completed a detailed report on impact mechanics at that location.

One difference which he observed between maars and meteorite craters was that maars were underlain by volcanic vents, while lenses of crushed and shattered rock occurred beneath meteorite craters. Another significant difference between the two forms was observed in their rims. He found the maar rims to be formed of thin layers of ash and other volcanic debris; and sometimes these layers were cross-bedded, making them deceptively similar to river-laid deposits. The rim deposits of meteorite craters, on the other hand, exhibited only relict bedding inherited from pre-existing stratified rocks. In the rims of both the Arizona Meteor Crater and the Odessa, Texas, crater he found that the original surface rock at the outer edge of the crater area had been overturned outward or buckled. At the Arizona crater, where rock layers of contrasting color had been penetrated, individual ejected blocks were overturned and the ejecta deposited in inverted stratigraphical order. Important differences existed between the rims of the two craters. As described by C. T. Hardy in 1953, "a

Moenkopi formation                                                              1
Kaibab limestone
Coconino sandstone and Toroweap formation

Fused rock                                                                      2

Shock front — Fused rock                                                        3

Shock front                                                                     4

Shock front                                                                     5

Reflected rarefaction                                                          6
Shock front

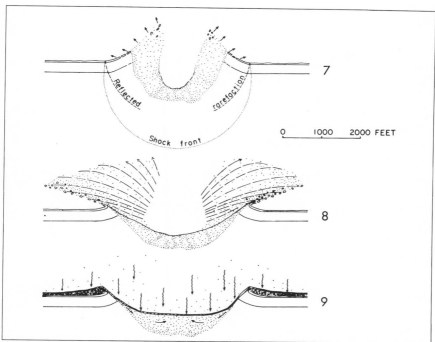

11.7    Diagrammatic sketches showing sequence of events in formation of Meteor Crater,
        Arizona. 1. Meteorite approaches ground at 15 km/sec. 2. Meteorite enters ground,
        compressing and fusing rocks ahead and flattening by compression and by lateral flow.
        Shock into meteorite reaches back side of meteorite. 3. Rarefaction wave is reflected
        back through meteorite, and meteorite is decompressed, but still moves at about 5
        km/sec into ground. Most of energy has been transferred to compressed fused rock ahead
        of meteorite. 4. Compressed slug of fused rock and trailing meteorite are deflected
        laterally along the path of penetration. Meteorite becomes liner of transient cavity.
        5. Shock propagates away from cavity, cavity expands, and fused and strongly shocked
        rock and meteoritic material are shot out in the moving mass behind the shock
        front. 6. Shell of breccia with mixed fragments and dispersed fused rock and meteoritic
        material is formed around cavity. Shock is reflected as rarefaction wave from surface of
        ground and momentum is trapped in material above cavity. 7. Shock and reflected
        rarefaction reach limit at which beds will be overturned. Material behind rarefaction is
        thrown out along ballistic trajectories. 8. Fragments thrown out of crater maintain
        approximate relative positions except for material thrown to great height. Shell of
        breccia with mixed meteoritic material and fused rock is sheared out along walls of
        crater; upper part of mixed breccia is ejected. 9. Fragments thrown out along low
        trajectories land and become stacked in an order inverted from the order in which they
        were ejected. Mixed breccia along walls of crater slumps back toward center of crater.
        Fragments thrown to great height shower down to form layer of mixed debris.

distinct asymmetrical anticline can be seen in trenches on all sides of the Odessa crater," while at the Arizona crater "downfaulted and downfolded blocks on both the north and south sides apparently constitute a graben-like structure."[22] In Hardy's opinion, these profound structural differences indicated that the two craters could not have been formed by the same process.

Shoemaker, however, observed that craters formed by underground nuclear explosions in the alluvium of Yucca Flat, Nevada, exhibited similar differences in rim structure. In a comparison of rims, he found in a nuclear crater known as Teapot Ess the same outward overturning of surface rock that he had observed in the rim of Meteor Crater. In the rim of another crater, called Jangle U, the rock was buckled in an anticline similar to that of the Odessa crater. These craters had been formed by nuclear explosions which centered at different depths.

Over the years, many theories had been advanced in attempts to estimate the size and velocity of impacting meteorites and the mechanics of crater formation, but their results differed widely. Shoemaker, in 1963, attacked the problem in reverse, as he put it, by deducing the requirements for theory "from the geology of Meteor Crater and its structural similarity to a nuclear explosion crater."[23] In fact, he remarked, "nearly all the major structural features of Meteor Crater, Arizona, are reproduced in a crater in the alluvium of Yucca Flat, Nevada, formed by the underground explosion of a nuclear device."[24] Extrapolating after careful measurements and calculations, Shoemaker concluded that the center of the meteoritic explosion which created Meteor Crater must have been at a depth proportionally similar to that which created the Teapot Ess crater. The Odessa crater, on the other hand, was probably formed by a shallower explosion, of depth comparable proportionally to that which formed the Jangle U crater. All four craters were underlain by lenses of breccia.

Shoemaker's investigations brought him to the conclusion that the crater-forming event which occurs when a large meteorite strikes the earth is immensely complicated. In his opinion, to call such an event an explosion (implying comparison with, say, the bursting of an overheated boiler) is an oversimplification, to say the least. Although the size and speed of a colliding meteorite might indicate that tremendous amounts of energy would be instantaneously released on impact, the event would be complicated by a great many factors. The shape and composition of the meteorite and of the terrestrial material it encountered would be of great importance. The extent of melting and vaporization of both terrestrial and meteoritic material would depend upon the various temperatures and pressures at which such changes took place in those materials—that is, upon their equations-of-state. At the moment of impact there would be intense compression of both the meteorite and the earth which it struck. A tremendous shock wave would course through the meteorite, and some—perhaps most—of its components would be melted or vaporized in the following rarefaction. A similar shock wave would course through the terrestrial rock, causing it to behave momentarily like a fluid[25]; in fact, as R. L. Bjork pointed out in 1961, the process is hydrodynamic in nature since the pressures generated are so much greater than the strengths of the materials. Outside the central region of vaporization and melting, the shock wave would create a zone of intense brecciation, and, at a great distance from the center, a zone of fractured rock. Thus, the conclusions reached by Shoemaker and by Beals, Innes, and Rottenberg were remarkably similar.

## CHAPTER 12

# More Controversy, More Craters

THE INVESTIGATION OF VARIOUS craters and curious geologic structures during the 1950s, and the conclusion that some of them had been created by meteoritic impact, attracted the attention of many scientists. Some geologists, however, remained skeptical of what they felt to be an increasing and unwarranted tendency to interpret cryptovolcanic structures as scars of meteoritic impact, or "astroblemes." One of the most skeptical, perhaps, was Walter H. Bucher, whose 1933 list of cryptovolcanic structures first drew attention to the fact that these peculiar formations exist in the United States. In 1963 Bucher, who was then professor emeritus at Columbia University and a widely respected "elder statesman" in geology, published articles in *Nature* and in the *American Journal of Science*, stating his reasons for believing that structures such as the Ries Basin in Germany, the Wells Creek basin in the United States, and the Vredefort dome in South Africa were not caused by meteoritic impact but by action from within the earth—that they were "geoblemes" rather than "astroblemes."

In general there was no disagreement as to the explosive origin of these structures. Controversy centered on the cause of the explosions. Bucher noted R. S. Dietz's proposal for the use of the neutral term "cryptoexplosion" rather than "cryptovolcanic" because of the uncertainties involved, and he suggested, perhaps with a touch of humor, that

> correspondingly, open craters without central uplift, from which no volcanic rocks have been ejected, should be called explosion craters (e.g., the Barringer Crater) to keep the reader's mind open to the possibility that they may not be the product of meteorite impact after all.[1]

Bucher remarked that until the end of the last century no features suggesting meteoritic impact had been found on the earth. Since then an increasing number of circular depressions on the earth's surface—the first of which, he noted, was "Arizona's much-advertised 'meteor crater'"—were being interpreted as products of meteorite impact; and craters which, like the Barringer crater, "are morphologically indistinguishable from explosion craters [i.e., maars]... are being called 'meteor craters.'"[2] He noted with apparent approval that although Gilbert in the 1890s was at first attracted to a meteoritic hypothesis for the origin of the Arizona crater, he finally rejected it in favor of a steam (or cryptovolcanic) explosion.

Bucher pointed out that if cryptovolcanic structures were indeed the result of meteoritic impact, they must be randomly oriented, with no systematic relation to any geologic structures. They must be independent of any past or present regional magmatic activity, that is, of subsurface movement of molten rock. And they must be "structurally of a nature that is comprehensible in terms of instantaneous impact and its immediate consequences."[3]

In his meticulously detailed study of the Ries Basin, Bucher showed that in his opinion these definitive criteria were not met. There was no lack of evidence of an explosion; it could be clearly seen in the uneven polishing and grooving which in places formed a "veritable pseudo-glaciated surface" and on which the direction of movement always pointed outward from the center of the basin. This had been observed early in the century by Branca and Fraas. But in Bucher's view the location of the Ries Basin could hardly be described as random. It was aligned with both the Steinheim Basin and with the Urach volcanic area, a remarkable cluster of what he called "volcanic tuff pipes." And this alignment lay parallel to a great crustal flexure; that is, a ridge in the underlying bedrock. He remarked:

> It would be difficult to find a more clearly defined position within the structural framework of southern Germany for a giant meteorite to hit. Pure chance could, of course, produce such an event. If this were a unique phenomenon, the argument involving chance would have little weight. But the meteorite hypothesis demands that the Ries giant had a small companion which is supposed to have produced the Steinheim Basin, and that it was placed neatly on the line connecting the Ries with the volcanic Urach area and, more remarkably still, almost halfway between these two major structures.[4]

And, in addition, Bucher noted that the Ries Basin lay at the precise spot where a broad anticlinal axis, or structural ridge, joined the crustal flexure.

With respect to his second point, Bucher noted that magnetic studies indicated the possible existence of basic intrusions (that is, rock containing iron and magnesium, which was in a partly molten state when it spread, or intruded itself, into the subsurface region) far below the ground surface in the general location of the Ries Basin. While most of the suevite tuff was magnetically neutral, magnetic anomalies discovered in some parts of it also suggested the influence of basic rock at depth. Another possible indication of magma below the ground surface was a high thermal gradient in a group of wells that had been drilled within the basin. And finally, with respect to his third point, the possibility that the structure had been created by instantaneous impact was completely ruled out by what he interpreted as a series of emplacements of suevite tuff: differences in its composition from one location to another within the Ries structure provided "indisputable evidence of successive eruptions."[5]

Coesite, discovered in the suevite of the Ries Basin in 1961, could have been formed, in Bucher's opinion, during violent expulsion of gas-laden particles through narrow vents. Shatter cones had been found in the Steinheim Basin, but not (at that time) in the Ries. He noted that Dietz had suggested that shatter cones found in many of the American cryptovolcanic structures might be evidence of shock from above; thus, "by a bold jump in logic," Bucher remarked, "he suggested that shatter cones may be diagnostic of meteoritic impact."[6] In Bucher's opinion, shatter cones were probably the result of mechanical shattering caused by gases at high pressure.

Bucher noted that cryptovolcanic structures in the United States and the Vredefort dome in South Africa occurred in locations similar to that of the Ries Basin, on large anticlinal flexures. They and the so-called satellite structures were in many cases arranged along these flexures in such a manner that they were aligned with "tuff pipes" indicating the presence of deep-seated active magmas from which gases in explosive quantities once arose. Because of all these considerations, Bucher rejected the meteoritic hypothesis for the origin of cryptovolcanic structures:

> I cannot accept a hypothesis that in two continents multiple meteorites (in one case a giant and a satellite, and in the other a giant and dwarf fragments) struck a clearly defined terrestrial flexure zone so that their impact scars are aligned parallel to its axis with structures of proved volcanic origin.[7]

A reply to Bucher's paper by R. S. Dietz appeared in the same (summer 1963) issue of the *American Journal of Science*. Dietz was a considerably younger man than Bucher, and he had, as he said, "espoused the case for impact structures for nearly two decades."[8] However, his reply, with the stated purpose not of argument but of discussing a few of the many questions raised by Bucher against the interpretation of cryptovolcanic structures as ancient meteorite scars, was gentle rather than polemical. While Dietz admitted that no final proof had yet been achieved, he nevertheless regarded most of the cryptovolcanic structures listed by Bucher in 1933 as astroblemes. He considered the Barringer Crater to be unquestionably of meteoritic origin and felt that "criteria for astroblemes may validly be extrapolated from it."[9] Although this crater lies in the midst of a great volcanic district—the volcanic San Francisco Peaks to the northwest; the White Mountains, peppered with volcanic craterlets, to the southeast; and a world-famous group of diatremes, the Hopi Buttes, to the northeast—he interpreted its location as an example of the random placement of meteorite craters. He remarked that "landing amidst this full span of volcanic effects was a most confusing thing for a meteorite to do but, with the perversity of nature, it apparently did so anyway."[10]

Barringer Crater, he pointed out, was by then almost universally accepted as a meteorite crater. As to the location of other structures, he remarked that it would be difficult to find a spot on the tectonic map of the United States that was not associated with regional trends; but "if we consider all of the cryptovolcanic structures," he said, "they seem to be randomly disposed."[11] He noted also that the ancient volcanic tuff pipes suggested by Bucher to be a source of explosive gas were, in many cases, of a geologic age vastly different from that suggested for the nearby cryptovolcanic structures. Dietz concluded by concurring with Bucher in that "traditional thinking should not be overlooked; we should not forego the mundane in favor of the esoteric."[12]

Bucher was an outstandingly able and energetic geologist. His papers were fine examples of traditional thinking—that is, of the careful interpretation of structural facts in the light of known principles. But his interpretation of the Ries Basin was made with no knowledge of the geological effects of hypervelocity impact, which were only beginning to be understood at that time.

Early the following year (February 1964), after reading an account of exploratory work at the Holleford crater, Bucher wrote a long letter to the author, C. S. Beals. He concluded with the following paragraph:

> Please forgive the length of this letter. But I need help in a sincere attempt to find a basis for distinguishing craters made by meteoritic impact from those that were formed by explosion from within the earth. . . . Yours is an important example that deserves the attention from the geologist's viewpoint that you have given it from the physicist's.[13]

Beals replied, in part:

> . . . I am heartily in agreement with your suggestion . . . that the great need at the present time is to find criteria for distin-

guishing unambiguously between craters formed by meteoritic impact and those formed by explosions from within the earth. It appears that this is not as easy as was first supposed, and I feel that it is going to be a long time before some of the most recent problems represented, for example, by features like the Ries of Nördlingen, the Clearwater Lakes, and the cryptovolcanic structures of the United States will be solved.[14]

A few months later, in May 1964, Bucher arranged to visit the Arizona Meteor Crater with a former student, Wolfgang Elston, a professor of geology at the University of New Mexico. Some years earlier Bucher had suffered an illness, and the exertions of field work were no longer easy for him. Nevertheless, he was eager for the trip, and in a letter to Elston he remarked that

> there are a number of specific things that we want to look for. . . . To me, the most important is to have a good look at the lower 6 feet of ejecta that form the rim. From my last visit I have the impression that no fragments larger than coarse sand size exist in those lowest levels. . . . Other items are structural details such as the flap of Moenkopi shale supposed to be bent back by the explosion, the nature of the "thrust fault" and of the vertical faults, and especially the nature of the allogenic breccia. . . .[15]

After a delay due to car trouble, the two men arrived at the crater, supplied with geologic hammers and Mason jars (Bucher's preferred container for rock samples), which he had managed to obtain from the depths of out-of-season supplies at a local store. They were met by E. M. Shoemaker, then Branch Chief of the U.S.G.S. Branch of Astrogeologic Studies, Shoemaker's eleven-year-old son, and Jack McCauley, a former student assistant of Bucher's at Columbia. In the parking lot of the Visitor Center, Shoemaker pointed out examples of shocked rock and explained his ideas concerning the crater; and Bucher, who was an excellent field geologist and an honest man, conceded the impact origin of Meteor Crater even before the tour began.

As mentioned in his letter, there were specific things Bucher wished to observe as the group set out to examine the crater. During a brief visit in 1938, he had noticed what is known as "reverse sorting," that is, coarse material lying above fine material, in rim deposits which Nininger described as ejecta. What, he wondered, had held the large pieces of rock in the air while the small pieces were raining down? Shoemaker explained that rather than ejecta, he had mapped these deposits as water-deposited sediments (alluvium), derived in part from a layer of mixed debris that initially blanketed the crater. The alluvial deposits were formed during the period when the crater contained a lake and now occur as terraces on the flanks of the crater rim.

Bucher knew that the Kaibab formation at the Grand Canyon was limestone, but he was not previously aware of a facies change from limestone to dolomite between the Grand Canyon and Meteor Crater. Thus, he had thought that the blocks of dolomite observed around the crater might have come from an older and deeper dolomite formation below the Kaibab, perhaps indicating some deep-seated disturbance. But at the crater he observed Permian fossils in the ejected dolomite blocks, identifying them as Kaibab.

Shoemaker pointed out the overturned "flap" of the Moenkopi and the thrust faults. They examined the crushed Coconino sandstone on the south rim and collected impactite droplets during the rest of the day. As they left, Bucher remarked, "Ah, but the Ries—that's still different!"[16]

In a letter to Elston written soon after his return home, Bucher said, "I have had few days in life so thoroughly crammed full of new insights into geological matters . . . I shall always treasure our trip as a rare bit of living."[17] Bucher's health was failing, and unfortunately no account of his new insights was published. He died less than a year later, in February 1965. After his death, a Walter Bucher medal was established by the American Geophysical Union, and an early recipient was R. S. Dietz. "Uncle Walt would have approved," Elston commented. "He loved good arguments, but never held rancor against his opponents."[18]

In Canada the search for meteorite craters continued. Summing up the results, C. S. Beals and his associates considered the most important advance up to that time (1960) to be "the establishment, with a high degree of probability, of the existence of three fossil craters of large size and sufficiently great age to justify the belief that others are likely to be found if a sufficiently exhaustive search is made for them."[19] The three craters referred to were the Brent and Holleford craters in Ontario, and Deep Bay in Saskatchewan. In addition they listed the following unexplained circular features as probable sites of meteoritic impact:

> West Hawk Lake, Manitoba
> Lac Couture, Quebec
> Manicouagan Lake, Quebec
> Clearwater Lakes (two craters), Quebec
> Carswell Lake, Quebec
> Keeley Lake, Quebec
> Ungava Bay, Quebec
> Hudson Bay arc
> Gulf of St. Lawrence arc
> five "stratified circular structures" in northeast Quebec
>     and Labrador

There was great curiosity about these structures, and eagerness to investigate them was intensified by the controversial nature of the whole inquiry—for many people were still skeptical of any meteoritic interpretation of terrestrial landforms.

The opinion long held by Bucher that the alignment of cryptovolcanic structures with geologic features of the earth provided evidence that they had been created by terrestrial processes was forcefully expressed in his 1963 paper, and also in a paper which appeared in 1965 after his death. And this opinion was shared by many geologists. For instance, G. H. J. McCall of the University of Western Australia, who was greatly interested in problems of both lunar and terrestrial cratering, noted that various North American cryptovolcanic structures described by Bucher in 1933 occurred in a narrow belt which apparently followed the trend of the Appalachian Mountain chain. Moreover, he pointed out that the fossil craters discovered in Ontario—which as he said had been "hailed as an addition to the astrobleme record"—were situated at the northern end of Bucher's cryptovolcanic belt. What McCall described as the manifest parallelism of this belt to a region of folding in the earth's crust left no doubt in his mind that the cryptovolcanic structures were created by processes originating within the earth. He remarked that similar arguments could be made concerning the Vredefort dome in South Africa. He felt, in fact, that the shatter cones which had been discovered at Vredefort by R. B. Hargraves in 1960 (see Chapter 14) should be viewed as indicating not that the Vredefort structure was created by impact, but that shatter cones could be created by recognized geological forces. As for coesite, he remarked that "this dense polymorph of silica seems, considering Bucher's devastating arguments, to have been erroneously given the virtual status of an impact explosion indicator."[20] Its discovery in the Ries Basin (which he considered to be volcanic) should, he felt, be interpreted as showing that it had been formed by processes originating within the earth.

Somewhat similar views were expressed by F. G. Snyder and P. E. Gerdemann in 1965 in their description of an alignment of eight distinct geological features in which volcanic eruptions, subsurface intrusions of igneous rock, and intense local deformation had taken place in the geologic past. These features lay in a structural zone which stretched more than 400 miles in an east-west direction from Illinois to Kansas. Hicks Dome, at the eastern end, was interpreted as a region where a deep-seated intrusion of magma occurred in late Precambrian time, resulting in the formation of breccia pipes and changes in the mineral content of local rocks. In the Avon region, a cluster of approximately eighty diatremes within an area of 100 square miles was similarly attributed to intrusions of magma in the geologic past. (It is interesting to note, incidentally, that these two features were considered by Bucher to be aligned on a curving axis with the Wells Creek basin. [21]) Other volcanic features in this east-west alignment were Furnace Creek and Hazel Green, where evidence of volcanic eruptions in the distant past could be observed in local rocks. In the Weaubleau region, intense faulting and brecciation probably indicated some ancient subterranean disturbance. At the extreme western end of the alignment, in the Rose Dome area of Kansas, several rounded hills had apparently originated when ancient granite intruded the subterranean rock, causing local elevations and intense faulting. Aligned with these features were the Crooked Creek and Decaturville structures, which had been listed by Bucher in 1933 as cryptovolcanic. Snyder and Gerdemann remarked:

> considered as isolated, unrelated phenomena, the individual features along this structural axis have been interpreted as cryptovolcanic explosions, meteoritic impact scars, igneous intrusives, and complex fault structures. Details of structure and lithology, subsurface information from core drilling, the remarkable alignment of the features, and the frequent association with basic igneous rocks suggest that these structures are closely related in mode of origin and that they represent intermittent deep-seated faulting and intrusion through a long period of time. [22]

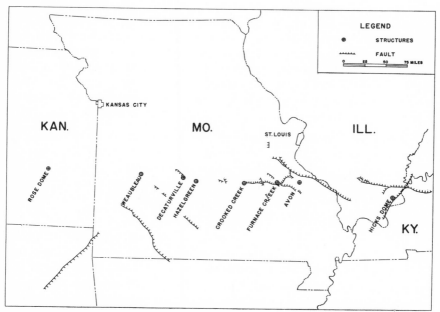

12.1 *Cryptoexplosion and volcanic structures, indicating explosive igneous activity along an Illinois-Missouri-Kansas axis.*

A different opinion with respect to the possibly meteoritic origin of certain terrestrial features was expressed by G. C. Amstutz in 1964. He felt that work with outer space was causing a "wave of meteorite-impact belief." He remarked that there had recently been what he described as "a mass psychosis of beliefs and visions of flying saucers," and that "what we now experience in regard to the revival of creationistic beliefs in impact from outer space is an alarming relapse into patterns which belong to the dark middle ages."[23]

In Canada the meteoritic origin of some of the recently discovered craters was questioned by K. L. Currie of the Geological Survey of Canada. After examining several craters, he reached conclusions that differed considerably from those of C. S. Beals and his associates, for in his opinion volcanism was clearly present in many of them. He remarked that the craters occurred in regions of uplift and suggested that they had been formed by collapse of the uplifted strata. "The characteristic cycle: upwarp, collapse, volcanism, is beautifully shown," he wrote. "None of these craters can be explained either by 'classical' impact theory or by analogy with known volcanic craters. New ideas are needed."[24]

The following year (1965) Currie published his observations about several Canadian craters. In his opinion, geophysical studies of Deep Bay, Saskatchewan, and West Hawk Lake, Manitoba, showed that they could conceivably be eroded meteorite scars, but in most other cases he found the arguments for meteoritic origin unconvincing. He interpreted the rim of the New Quebec crater as the remains of a collapsed dome and declared that some of the rocks of which it was composed had been chemically altered by processes related to volcanism. He felt that the location of the Brent crater on a major graben, and the presence of unusual rocks beneath it, indicated that it was created by forces from within the earth. Concerning the Holleford crater, he remarked that

> unless the presence of coesite is regarded as diagnostic, Holleford can hardly be regarded as a proven impact crater. The shape is wrong, the supposedly impacted rocks do not show the expected fracture pattern, and the structural relations are very suspicious. Judgment should be suspended until more data are available.[25]

Currie noted that the four very large craters (the two Clearwater craters, Carswell, and Manicouagan) in which central uplifts had been observed exhibited a form which was different from that of the smaller craters, being for one thing proportionally shallower. He suggested that the large craters were probably the result of a geologic sequence in which updoming was followed by volcanic eruptions and subsidence, then by extrusion of lava, and finally by uplift which

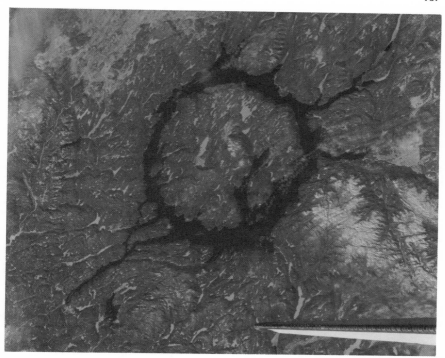

12.2   The Manicouagan meteorite crater, with a diameter of about thirty-nine miles (65 km).

affected only the central parts of the craters. "There is no logical point," he remarked, "at which a meteorite can be interpolated into this sequence."[26] He suggested also that certain small craters such as Merewether and Lac Chatelain (another small crater in northern Quebec) might have been formed by the collapse of mounds caused by frost upheaval. (Such a mound is called a pingo.) Finally, the fact that the Canadian craters apparently occurred along lines of uplift in the Canadian Shield offered conclusive evidence, in his opinion, that they were the result of forces connected with the gradual evolution of the earth.

These widely diverging views held by competent geologists give some indication of the enormous difficulties which were involved in the exploration and interpretation of the various craters. Nevertheless, the inevitable and ongoing controversy which resulted was a stimulant to further investigation. In fact, investigation continued with remarkable vigor, considering the complex problems presented by distance, remote localities, rugged terrain, and transportation of equipment—to name a few; and the number of "probable" Canadian meteorite craters accepted by several outstanding authorities soon

increased to ten. As listed by I. Halliday and A. A. Griffin in 1964, and
by C. S. Beals and I. Halliday in 1965, they were, in order of increasing size:

|  | Diameter in kilometers |
|---|---|
| Holleford, Ontario | 2 |
| Brent, Ontario | 3 |
| New Quebec, Quebec | 3 |
| West Hawk Lake, Manitoba | 4 |
| Deep Bay, Saskatchewan | 12 |
| Lac Couture, Quebec | 14 |
| East Clearwater, Quebec | 22 |
| Carswell, Saskatchewan | 32 |
| West Clearwater, Quebec | 32 |
| Manicouagan, Quebec | 60 plus |

Other locations suggested a year or two earlier by C. S. Beals *et al.*
were felt to be doubtful. Keeley Lake and Ungava Bay had not been
investigated, nor had the five stratified circular structures. Seismic
methods had been used by P. L. Willmore and A. E. Scheidegger in
1956 to examine the sea bottom in what was known as the Gulf of St.

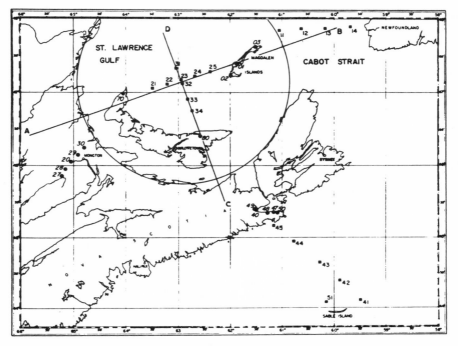

*12.3   Area of operations and position of shot points and recording stations during seismic
examination of the sea bottom in the Gulf of St. Lawrence arc.*

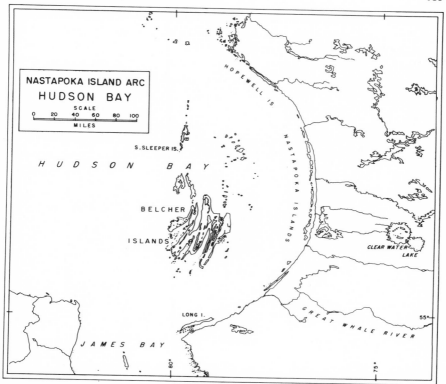

**12.4** *The Nastapoka arc on the east coast of Hudson Bay. The circular arc is a conspicuous feature, suggesting an unusual origin. A resemblance to Mare Crisium on the moon has been suggested.*

Lawrence arc, a remarkably circular sweep of shoreline formed by the northern coasts of Cape Breton Island and Nova Scotia, and the eastern coast of New Brunswick. Results were inconclusive. Seismic observations, however, indicated sediments within the arc to a depth of almost four miles, and the investigators determined the existence of what they described as a cavity in the crust, the dimensions of which, in their opinion, were comparable to those which would be expected from a meteorite explosion.

Various observers had drawn attention to the Hudson Bay arc, also called the Nastapoka arc, a surprisingly round indentation in the east coast of Hudson Bay, where a long, thin line of near-shore islands followed much of the curving coastline. Some 400 miles long, its shape was that of an arc of a huge circle. Hudson Bay was known to be a shallow sea, with low and featureless coastlines to the south and west. To the east a deep basin lay within this arc, and its shoreline was edged by rugged hills. Altogether the large area of Hudson Bay, the great age of the surrounding rocks, and the complexity of what little was known of the regional substructure suggested, as C. S. Beals remarked, "a problem of great difficulty, and at the same time of exceptional interest."[27] As Beals pointed out, a strong argument in favor of impact origin was that it was difficult to imagine any other process that might produce a perfect circle of such great size.

I. Halliday noted that craters similar to the Hudson Bay arc in their probable great age and large size had been observed on the moon and might therefore be reasonably expected to occur on the earth. J. Tuzo Wilson, a geologist of wide interests, welcomed Beals's suggestion of a possibly meteoritic origin for the feature. He felt that Beals's work

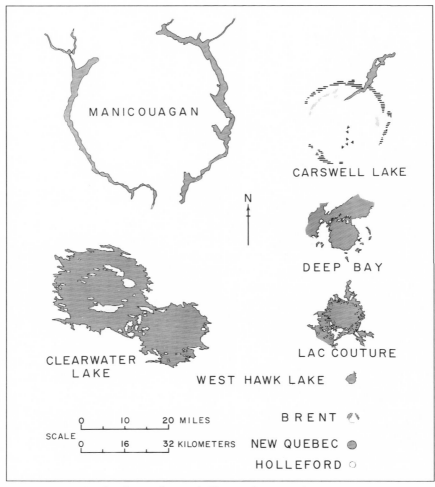

*12.5   Comparative sketch map of ten probable Canadian meteorite craters.*

"emphasized a new geological process and thus broadened and gave scope to ideas about the earth."[28] He pointed out that if the Hudson Bay arc had indeed been created by meteoritic impact, the event probably occurred a billion or so years ago, and that the resulting crater had no doubt been so altered by processes of erosion, and possibly even by the movement of continents, that few criteria necessary to resolve the problem were likely to be found.

The relative sizes of the ten Canadian craters of "probable" meteoritic origin were graphically shown by M. R. Dence. Gravity, magnetic, and seismic surveys had been carried out in all these locations, along with geological and structural examinations, and in some cases the findings were supplemented by diamond drilling. Some of the basic criteria for the recognition of meteorite craters, such as circular plan and underlying breccia, were observed in all ten examples. The smaller craters, those ranging in size from Holleford (diameter slightly more than a mile, or two kilometers) to Lac Couture (diameter eight miles, or 14 kilometers), could unquestionably be classed as explosion craters. The four larger structures, however, with diameters ranging from twelve to more than thirty-five miles (20 to more than 60 kilometers), exhibited some peculiarities. Dence called them "complex" craters and remarked that their identification as meteorite craters was based on the fact that, besides their differences, they displayed many similarities to the smaller, or "simple" craters.

Striking similarities, in fact, existed between individual large and small craters, in spite of the central uplifts to be found in the complex craters, the concentric ridges and troughs which encircled some of them, and the fact that with increasing diameter they were relatively shallower. The Deep Bay crater, for instance, one of the least eroded of the simple craters, was surrounded by a system of concentric fractures which enclosed a circular area with a diameter of about twice that of the crater itself.[29] Part of an apparently similar fracture ring, in a much more advanced stage of erosion, had been observed at the Brent crater by Millman in 1960. These circular fractures suggested an affinity with the concentric ridges and depressions encircling some of the larger craters.

Rocks in the central uplifts that existed in some of the craters displayed puzzling features. In 1962 highly unusual metamorphic rocks were found on islands of the Clearwater Lakes by D. B. McIntyre. Among other deformations of minerals contained in these rocks were feldspars "recrystallized in sheaf-like aggregates," and microbreccias having "numerous deformation lamellas [i.e., thin layers] in the quartz."[30] During the winter of 1964, diamond drilling of the two Clearwater craters was undertaken from the frozen surface of the lakes. East Clearwater Lake had at first appeared to be a much-eroded simple crater. Drilling showed it to be considerably shallower than expected. It also revealed a pronounced central uplift, almost obscured by sedimentation and glacial erosion, where deformed gneiss had been raised far above its topographic position in the surrounding region. Fine shattering, intense fracturing, and general deformation and alteration of the rock structure were found in the uplift. In the larger West Clearwater Lake it was found that the central ring of islands was a circle of promontories on the outer edge of a broad central uplift. Most of the uplift, which was composed of fractured and deformed gneiss, was under water.

There was a central uplift at Carswell Lake, too, more eroded than that of West Clearwater, but similar in size. In the opinion of some investigators, the Carswell Lake structure was cryptovolcanic. It was unique among known Canadian craters as it occurred not in gneiss but in a thick sequence of originally flat-lying sediments. During the formation of the central uplift, underlying Precambrian granite had evidently been raised several hundred feet above its normal position in the region, and the structure was encircled by a ring-shaped trough in the surface sandstone. Dence pointed out its similarities to the Clearwater craters and noted that besides presenting structural

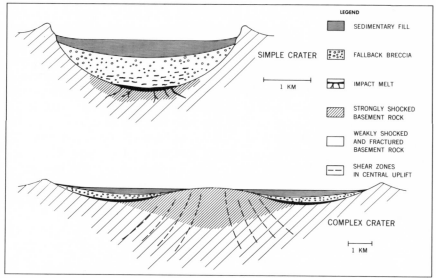

12.6   Schematic cross sections of simple and complex meteorite impact craters. Simple craters (upper) *such as Brent, Ontario, exhibit an original depth that may be as much as one quarter to one third the diameter and are partly filled by fallback breccia. In complex craters* (lower), *such as Clearwater Lakes, Quebec, inward and upward movement of the rocks below the crater in response to the impact produces a central uplift and results in a structure that is much shallower in comparison to its diameter. In simple craters, fallback breccia and impact melt are concentrated toward the center. In complex craters, these deposits are thickest in an annulus which surrounds the central uplift.*

problems similar to those observed by R. B. Hargraves in 1961 in the Vredefort Ring in South Africa, it was similar to the cryptovolcanic localities described by Bucher in 1933.

In the largest of the "probable" Canadian craters, Manicouagan, almost forty miles (65 km) in diameter, a central uplift of crushed and altered gneissic rocks rose more than 1,000 feet (300 m) above the surrounding breccia, and the structure was encircled by concentric, ring-shaped ridges and depressions. Similar origins for all these craters seemed indicated. Dence remarked, however, that in order to explain the complex crater form, major extensions of the present theory of impact mechanics would be necessary.[31] Later (1968) he noted that studies at various impact craters indicated that a decrease in the intensity of metamorphic effects occurred radially outward from the center. As rocks in the central uplifts of complex craters exhibited low to moderate grades of shock metamorphism, he suggested that differential pressures created by the impact caused the rocks below the crater to be drawn upward into the center.[32] The central uplifts are now considered analogous to the central peaks of lunar craters.

As time went on, new discoveries lengthened the list of Canadian impact sites. One of them, La Malbaie, on the north shore of the St. Lawrence River, was found in an unusual manner. The first indication that it might be an impact site was provided by the presence of shatter cones, discovered by J. Rondot in 1967.[33] Investigation revealed the characteristics of a complex crater between La Malbaie on the northeast and Baie St. Paul on the southwest—complete with peripheral depression and central uplift, but with the southeastern half of the structure cut off by the St. Lawrence River. Impact origin was confirmed by various evidences of shock (which will be discussed in the next chapter)—kink bands in biotite, deformation of quartz grains, and impact melts. As P. B. Robertson remarked, "this marks the first instance of the initial discovery of a meteorite crater by the finding of shatter cones, and as such, perhaps vindicates the proponents of shatter-coning as an undeniable criterion for the identification of impact craters (Dietz, 1960)."[34]

CHAPTER 13

# Impact Metamorphism

IN THEIR EXAMINATION OF cores obtained by diamond drilling at the Brent crater, C. S. Beals *et al.* reported many indications that the shattered rock in the breccia lens below the crater floor had been subjected to great heat. Fragments of gneiss and granite had evidently undergone deformation and partial melting and in some cases were embedded in a glassy matrix. In addition, within a more or less central region of the crater they found a layer of once-melted material which they called a lava sill; that is, a thin layer of intrusive igneous rock. Similar lava-like material was found in a number of other Canadian craters and was thought by some investigators to be an evidence of volcanism. In spite of its volcanic appearance, however, this layer was interpreted by M. R. Dence as indicating that, as a result of great heat, the fragmented rock had been fused into a glassy mass. Rather than an intrusive layer, he considered it to be an integral part of the melt zone, perhaps marking the limit of penetration by the meteorite. Dence also noted that partially melted and deformed rock fragments from the Brent crater had been compared with samples of breccia taken from several ancient diatremes. No trace of the peculiar type of deformation found at Brent was discovered in the diatreme breccia.[1]

Microscopic examination of deformed, altered, and fused rock taken from the various Canadian craters revealed features that were quite different from those of any known volcanic rocks. Quartz grains taken from the meteorite craters, for instance, often contained clusters of tiny parallel lines which gave the grains a layered appearance. Various kinds of glass had been formed by the melting of quartz, feldspar, and other components of the rock. In some samples different kinds of glass were found close together in the form of drops and threads, apparently indicating that the moment of intense heat which formed them was so brief that the melts had cooled and congealed before mixing could take place.

Somewhat similar observations were made by investigators of other craters, the earliest of which were no doubt those of Barringer and Tilghman in their examination of Meteor Crater. Merrill, in 1907, described what he called "a peculiar and very local metamorphism of this rock." Under the miscroscope he found that some samples exhibited features "which are quite at variance with our ideas of the stable character of quartz sand." He noted also ". . . in the still firm rock a large number of granules which are so completed changed as to give rise to forms at first glance scarcely recognizable as quartzes at all."[2] It was during this investigation that he noted the quartz glass, similar to fulgurite glass (which, incidentally, was not known as lechatelierite until 1915, when it was so named by Lacroix) and which he cited as evidence of heat much greater than that likely to occur in volcanic processes. But scientific opinion had changed since the early years of the century, and the origin of Meteor Crater was no longer a matter of speculation. Merrill's observations concerning the peculiarities of the Coconino sandstone to be found there were recalled when similarly distorted quartz grains were found in other craters of suspected meteoritic origin, and their peculiarities were attributed to the high pressures created by meteoritic impact. Effects of high pressure became a matter of intense interest, and attention turned to results of high-pressure research in other fields.

Certain minerals were known to be deformed when subjected to the great pressures involved in natural processes of metamorphism and rock flow, but the mechanics of such deformation was a long-standing geological problem. And a relatively new problem that stirred up tremendous interest in the early 1960s had to do with the effects of shock pressures on various materials. Because these investigations, some of which had been going on for many decades, were reaching conclusions of great importance to the investigators of meteorite craters, let us digress briefly to see how they came about.

Investigation of the effects of pressure began long ago. In the seventeenth century Robert Boyle showed that the volume of a gas is inversely proportional to pressure. The fact that pressure affects both the boiling and freezing temperatures of water was also known very early. It was not until the nineteenth century, however, that numerous investigators carried out experiments designed to determine the effects of high pressure on optical properties, phase changes, electrical resistance, and other characteristics of various materials. Such experiments involved two major difficulties. One was the creation of the high pressure. The other was the enclosing of the material to be compressed in containers that would be capable of withstanding the pressures involved. An early investigator was Jacob Perkins, who in 1820, as P. W. Bridgman has remarked, cast his experiments in a mold heroic for those days: he used a cannon as the containing vessel and obtained high pressures by sinking his apparatus to known depths beneath the sea.[3] By the early decades of the twentieth century, technical ingenuity had increased enormously. Between the work of Perkins and the development of techniques which made it possible to reach without leakage any pressure allowed by the mechanical strength of the container lay more than a century of experimentation.

As time went on, high-pressure research was extended partly because of purely scientific interest and partly because of industrial needs for new materials and for information concerning effects of high pressure on metals and alloys. After World War II, additional impetus came from the U.S. Air Force and from the developing space program,[4] both of which required materials able to withstand very high pressures and temperatures.

Besides many practical applications, high-pressure research contributed substantially to developments in various sciences. Prominent among them were solid-state physics, concerned with the behavior of solid material in general, and several branches of geology, concerned with the behavior of rocks and minerals in particular. Early in this century seismic studies showed that a central core of extremely dense material exists within the earth, and gravitational pressures surrounding it were calculated to be enormous. Continuing seismic explorations revealed the existence within the earth of several discontinuities—that is, regions where at fairly uniform depths seismic waves are refracted. Did these discontinuities indicate changes of material, or did they perhaps indicate changes of phase? It was thought that materials common on the surface of the earth might assume quite different forms when subjected to the enormous pressures existing at depth. Speculation about the earth's interior increased interest in the composition of various rocks, especially those suspected to have formed under pressure far below the surface. In fact, Francis Birch remarked in 1963 that "it is hardly an exaggeration to say that every aspect of geophysics and geochemistry of the earth's interior requires consideration of the effects of pressure."[5]

Another group of scientists, the mineralogists, had long been interested in the effects of pressure on the formation of minerals. In regions subjected to the slow, heavy movement of one great mass of rock against another (the process known as regional metamorphism, which takes place at the elevated temperatures prevailing below the earth's surface) mineralogical changes were known to occur. Evidently, increased pressure affected the manner in which certain constituents of the rocks combine to form new minerals. A long-standing mineralogical problem had to do with the foliated, or lamellar, structure often observed in the minerals of deformed, or metamorphic, rocks. At the turn of the century, F. D. Adams and J. T. Nicolson undertook an experimental investigation of the flow of marble, noting in their review of earlier work that a widespread divergence of opinion existed as to how—or even if—the flowing of rock came about. They subjected samples of marble to conditions which simulated as closely as possible various natural deformational environments, the contributing factors of which, it was generally agreed, must be (1) great pressure (which must be differential, or directed pressure; mere "cubic compression," or confining pressure, though it might produce molecular rearrangement in the rock, would not cause flow); (2) high temperature; (3) percolating water; and (4) time. (Later investigators recognized the importance of both confining and differential pressure.)

They found in many cases that experimental deformation of marble, under some combination of these conditions, resulted in a foliated or laminar structure which they attributed to a mechanical flattening and elongation of the constituent calcite grains—a phenomenon which had apparently not been previously recorded. Movement of the calcite particles was compared to flow similarly induced in metals subjected to high pressures and temperatures, and which, as they pointed out, had been interpreted as due to distortion of metallic grains by slipping along "gliding planes" related to the crystal structure of the material. Adams and Nicolson then examined specimens of marble from widely separated folded, or metamorphosed, regions. In many of their samples they observed a foliated, or lamellar, structure which they interpreted as an effect of movement and which resembled similar effects in the artificially deformed rocks. Their conclusions, given in some detail, stated in part that under suitable conditions of pressure there can be a flow of marble just as there can be a flow of metal. Adams continued his investigations for several years. Although he left no doubt as to the existence of rock flow, the mechanics of the process was not well understood; but early in the century it was widely recognized that metamorphic rocks contain a preponderance of lamellar minerals.

In 1928 M. J. Buerger published an account of deformations that he had observed in crystals of certain ore minerals. He noted that the plastic deformation of crystals seemed to be a subject of little interest to American crystallographers and mineralogists and that investigations along such lines had been carried out "mainly by our more academic German contemporaries."[6] He remarked nevertheless that the deformation of crystals must inevitably become important to mineralogists because, in order to understand the plastic flow of rocks, an understanding of the plastic flow of crystals is essential.

In 1930 Bruno Sander published the results of many years of investigation in a volume called *Gefügekund der Gesteine.* This influential work, which was not widely known in the United States until it was introduced to English-speaking geologists in 1933 by Eleanora B. Knopf, has formed the basis of the modern science of structural petrology, or petrotectonics. (The more usual term today is petrofabrics.) Knopf pointed out that Sander stated, as a fundamental principle, that rocks are constituted in such a way that their minute components are symmetrically related to the forces by which the general structure of the rock mass is controlled; and, in addition, that he explained the orientation by shape of mineral grains within deformed rocks as resulting from their plastic deformation in response to compressive forces. "The ultimate test of this theory," Knopf remarked, "will be the study by petrotectonic methods of material that has been deformed under known stress conditions."[7]

Both the flattening of calcite grains in marble and the alignment of quartz grains in a definite crystallographic direction (known as "preferred orientation" and often observed in metamorphic rocks) could, according to Sander, be attributed to the type of deformation to which the containing rock had been subjected. The alignment of the quartz grains indicated, too, that under certain conditions quartz could be highly mobile. In an attempt to determine whether this apparent mobility indicated some sort of plastic deformation of individual grains of quartz, experiments were conducted by David Griggs in which known stress conditions were provided by means of a high-pressure apparatus devised by P. W. Bridgman. Under pressures considerably higher than those available to Adams and Nicolson at the turn of the century, calcite crystals were indeed deformed, but no deformation occurred in similarly treated quartz crystals. It was found, in fact, that under increasing pressure, quartz displayed increasing strength until, instead of undergoing deformation, it finally broke. As H. W. Fairbairn noted in 1939, quartz was something of an *enfant terrible,* and its peculiarities, which were related to its immensely strong, close-knit structure, had long been the despair of students of mineral deformation. David Griggs and J. F. Bell remarked in 1938 that "whether quartz is really plastic at high pressures and temperatures or whether the observed mobility is only an 'apparent plasticity' is as much of a moot question today as 50 years ago."[8]

Perhaps the interest and urgency felt by many geologists with respect to problems concerning the effects of high pressure on rocks—problems closely related to the mechanics of rock flow—was best expressed by the physicist P. W. Bridgman. He concluded his 1936 account of experiments undertaken to determine the effects of high pressure on various minerals with the following remark: "Geology is rapidly approaching a point where an ultimate problem which has

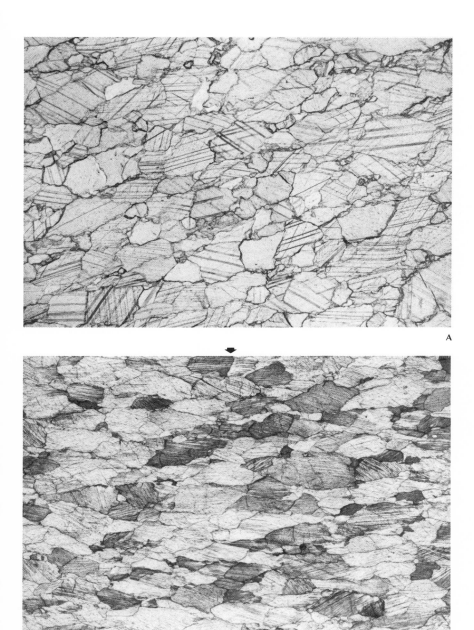

**13.1** *Photomicrographs of Yule marble. A: Undeformed Yule marble, cut perpendicular to planar layering (foliation) in rock. Plane polarized light. Cross-cutting lines in individual calcite crystals are twin planes, along which slip can occur (or has occurred). B: Yule marble experimentally deformed at 300°C and 5,000 atmospheres confining pressure. Plane polarized light. Arrows indicate direction of compression. Note elongation of calcite crystals perpendicular to compression direction. Deformation has caused calcite twin planes to become much finer and mainly perpendicular to compression direction.*

been staring it in the face can no longer be sidestepped, namely, to determine the actual specific physical and chemical behavior of those materials which do actually constitute the earth's crust."[9]

Stimulated by Sander's method of microscopic analysis of rock fabric, interest concerning the plastic deformation of crystals increased rapidly among mineralogists, as Buerger had predicted that it must. By the 1940s there was general agreement that experiments were necessary to determine the effect of high pressures and temperatures on various rocks and rock-forming minerals. In fact, a Committee on Experimental Deformation of Rocks was organized in 1944 by the National Research Council in order to plan a systematic approach to such work.

In the years immediately following World War II a great deal of experimental work was done, much of it by David Griggs and his associates. An extensively used test material was Yule marble from Gunnison County, Colorado, a rock composed entirely of calcite grains. Samples of Yule marble were subjected to various high pressures and temperatures.[10] It was found that after exposure to temperatures of 300° to 500°C and pressures of about 5,000 atmospheres, the resulting deformation closely resembled that observed in naturally deformed marble. It was suggested that the force by which a large mass of marble had been deformed in nature might be revealed through microscopic determination of the orientation of the constituent calcite grains. One condition necessary for the estimation of natural stress, however, was that the mechanism of deformation (the gliding mechanism) must be known. And in the mid-1950s, calcite and dolomite were the only rock-forming minerals for which this was definitely established.

Various investigators had observed that in deformed rocks containing both quartz and calcite, the orientation of the grains of the two minerals was so similar as to suggest that quartz, too, might be a reliable indicator of the stress to which the containing rocks had been subjected. After many unsuccessful attempts, the laboratory deformation of quartz was at last achieved in 1961 by N. L. Carter, J. M. Christie, and D. T. Griggs. Later, in 1964, these investigators described the artificially deformed quartz as indistinguishable from naturally deformed quartz in all respects except that of crystallographic orientation. As a possible explanation of this difference, it was suggested that time may have been a crucial factor. In natural deformational environments, the vast periods of geologic time during which the rocks were subjected to abnormally high pressures and temperatures may have compensated for the fact that the natural pressures and temperatures were lower than those used experimentally, and might account for the observed difference. The investigators concluded that the discovery that quartz may be deformed and recrystallized with ease in the laboratory and that the resulting material bears so much resemblance to naturally deformed quartz leads one to hope that further studies may reveal additional diagnostic features of lamellae and other structures which may aid in unravelling the complex tectonic history of rocks. [11]

In an attempt to realize this hope, a careful study of naturally occurring deformed calcite and quartz in calcite-cemented sandstone from Dry Creek Ridge Anticline (Montana) was undertaken by N. L. Carter and M. Friedman in 1965. They concluded that the orientation of stresses through which rock had been deformed could indeed be determined from microscopic examination of the quartz.

The mechanism of quartz deformation was of tremendous interest to the investigators of meteorite craters, for whom the solution of this long-standing problem was most fortunately timed. Stimulated by the ongoing controversy over craters (cryptovolcanic versus meteoritic origin) and by the peculiar samples of quartz found at sites of suspected meteoritic impact, N. L. Carter in 1965 made a study of quartz samples from Meteor Crater in Arizona (where peculiarities of quartz grains were noted by G. P. Merrill as early as 1907), from Clearwater Lakes in Quebec (where deformation of quartz lamellae was reported by D. B. McIntyre in 1962), and from the Vredefort ring in South Africa. He found that these samples were consistently similar to experimentally deformed quartz, and consistently different from quartz which had been deformed by ordinary geologic processes. Judging by the temperatures and pressures involved in experimental deformation, he calculated that the samples had been subjected to pressures of between 35,000 and 60,000 atmospheres, at temperatures between 700' and 1500'C. Such enormous pressures and temperatures—immensely greater than any that occur in volcanic events—could have been produced in nature on the surface of the earth only by the impact of an extraterrestrial body. Carter concluded, therefore, that the fact that quartz of impact structures resembles experimentally deformed quartz rather than the quartz of metamorphic rocks could provide another criterion—in addition to coesite, stishovite, and shatter cones—for the recognition of meteoritic impact.

While these investigations into the deformation of quartz were going on, research was continuing on various aspects of diamond formation. These researches, too, were of great interest to the investigators of meteorite craters. Writing in 1955 about the graphite-diamond equilibrium (that is, the temperature and pressure conditions under which the two phases of carbon exist together in a stable relationship), R. Berman and F. Simon remarked that

The problem of the synthesis of diamond has naturally occupied the minds of men for a considerable time. In most of
the attempts at synthesis it was hoped to produce conditions
under which diamond would be the stable form of carbon, but
it must be emphasized that this is not necessarily the only
possible way. Whatever the process envisaged, however, the
equilibrium diagram for the two carbon modifications is
bound to be a most valuable guide. . . . All authors have given
the same order of magnitude for the equilibrium pressures
and temperatures which would be high enough for the reaction to be appreciable. . . .[12]

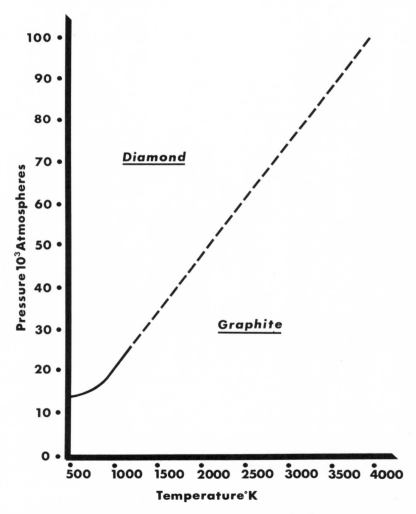

13.2   *The graphite-diamond equilibrium curve.* _____ *= calculated up to 1,200°K;* -------- *–
linear extrapolation*

And they concluded:

> While this paper was being written the General Electric Company of America published information about their successful experiments on the synthesis of diamond, which certainly represents a great technical feat. The pressures and temperatures at which the synthesis has been achieved are in good agreement with the values calculated here and make it probable that the synthesis was achieved under equilibrium conditions.[13]

Incidentally, it is interesting to note that under ordinary day-to-day conditions of temperature and pressure, diamonds are unstable. There is no perceptible reversion to graphite because of the infinitesimally slow rate of that reaction.

At the General Electric Laboratory in Schenectady, New York, where this was accomplished, some conditions peculiar to the formation of diamond were listed in 1959 by H. P. Bovenkerk and his associates. They remarked that all the observed cases of diamond preparation in their laboratory occurred at pressures and temperatures appropriate for the thermodynamic stability of diamond, but their work indicated that diamond could form in a number of ways and that stubborn mysteries still surrounded some of them. Several chemical systems were studied in connection with diamond synthesis, and the most successful were those involving carbon dissolved in various molten metals, which apparently acted as catalysts. The most effective metals were nickel and iron.

In 1956, as we have seen, Nininger suggested the possibility that diamonds in the Canyon Diablo meteorite had been created by the shock of impact. He also noted that the diamond-bearing specimens of Canyon Diablo iron (for diamonds did not occur in all samples) were always small, that their Widmanstätten patterns showed evidence of alteration by heat, and that they were found only on the rim of the crater.

Because of Urey's theory that meteoritic diamonds had formed deep within moon-sized asteroids, the origin of meteoritic diamonds was a matter of interest to astronomers as well as to geologists. Urey's ideas received considerable support, but arguments against the existence of large, diamond-bearing asteroids early in the history of the solar system continued to appear. In 1960 evidence published by G. G. Goles, R. A. Fish, and E. Anders suggested that parent bodies of meteorites, rather than being moon-sized, were probably no larger than about 250 kilometers in radius—too small to have provided pressures necessary for the formation of diamond. And strong support for Nininger's suggestion was given by M. E. Lipschutz and E. Anders in the same year. (Their paper, *The Origin of Diamonds in Iron Meteorites,* was presented orally in 1960 but was not published until 1961.)

Lipschutz and Anders pointed out that formation of the Widmanstätten pattern in nickel-iron requires long, slow cooling from high temperatures, and that metallurgical studies[14] of meteorites indicated complicated histories of melting, cooling, and reheating over immense periods of time. According to their calculations, bodies of lunar mass (such as those suggested by Urey as probable parent bodies of meteorites), if formed by cold accretion followed by radioactive heating to temperatures high enough for formation of diamond, would not have had time during the entire history of the solar system to cool to the relatively low temperature required for formation of the Widmanstätten pattern. The changes in these patterns, observed by Nininger in the diamond-bearing specimens, indicated that they had been subjected to high temperatures *after* the patterns had been formed. The fact that the changed patterns had remained imprinted on the specimens (instead of disappearing completely) indicated that the period of reheating had been extremely brief—Lipschutz and Anders estimated 1 to 5 seconds—and that it was followed by rapid cooling. Such rapid temperature changes could not take place in a body of large mass and must therefore indicate that the reheating and cooling occurred after the specimens had been reduced to their present small size; that is, after the final break-up of the Canyon Diablo meteorite.

In addition, Lipschutz and Anders noted that one of their samples of Canyon Diablo iron, a diamond-bearing nodule of troilite (a mineral of relatively low density) , indicated exposure to temperatures considerably higher than those to which the heavier minerals adjacent to it had apparently been subjected. They concluded that a thermal gradient such as was indicated by this phenomenon could not have been caused by an external heat source. It could have been caused only by a shock. (This was an example of what later became known as *selective phase transition,* a unique criterion of shock.) "If Nininger's correlation of re-heating and the occurrence of diamonds is valid," they remarked, "then the conclusion is inescapable that diamond formation in Canyon Diablo took place upon impact with the earth rather than pre-terrestrially."[15] They concluded also that the most severely shocked fragments had been ejected from the exploding meteorite at a relatively low velocity, and thus have been found only on the rim.

Anders remarked that the presentation of this evidence by Lipschutz in April 1960 at a meeting of the American Chemical Society "aroused deep apathy."[16] But interest increased when, only three months later, the discovery in the Arizona crater of the first natural occurrence of coesite provided empirical verification of the fact that a shock wave was created by the impacting meteorite. Shock pressures of at least 20,000 atmospheres had evidently existed long enough to form microscopic crystals of coesite. It therefore did not seem un-reasonable to suppose that pressures had been high enough for formation of diamond. They concluded that the diamonds in the Canyon Diablo iron were formed at the moment of impact either by direct solid-state transformation of graphite or by growth under shock pressure from an iron carbide, iron sulfide melt. Therefore, to regard meteoritic diamonds as evidence of the former existence of very large meteorite parent bodies, as affirmed by Urey and others, [17] no longer seemed justified. Further confirmation of these conclu-sions was noted by Anders in 1965 with metallurgical evidence that a moderately shocked specimen of Canyon Diablo iron indicated expo-sures that briefly exceeded 130,000 atmospheres.

A few months later, R. N. Wentorf, Jr., and H. P. Bovenkerk suggested that there had by then been enough experience in labora-tory production of diamond for useful comparisons to be made be-tween man-made and natural ones. Small black diamonds very simi-lar to those found in meteorites had been produced in the laboratory from carbon in the presence of molten metal, or from a mixture of the two, under conditions of relatively low temperature and moderately high pressure. Similar conditions could have existed within a large body far out in space, or during the "collision process." They felt, with respect to diamonds in stony meteorites, that more study was re-quired. With respect to Canyon Diablo diamonds, they found the arguments of Lipschutz and Anders convincing in many ways, except that the average diameter of the tiny diamonds found in meteorites was known to be about 0.01 millimeter. Lumps of diamond with diameters of more than 1 millimeter—a hundred times larger—had been observed in the Canyon Diablo iron. They felt that this provided a strong argument against impact origin because, judging by their experiments, a time of at least one second would have been required for the growth of such relatively large diamonds; but the duration of the required pressures at the moment of impact had been calculated at only one tenth of a second. Commenting on these conclusions, Anders noted that Wentorf and Bovenkerk had dismissed the possi-bility of shock-induced conversion of graphite to diamond. [18] He pointed out also that problems concerning the origin of diamonds in stony meteorites—whether formed under equilibrium or shock conditions—might be resolved if such diamonds showed preferred orientation, which would provide evidence of shock.

A shock wave is an extremely strong pressure wave that travels with greater velocity than the speed of sound. Investigation of the effects of shock was accelerated during the 1950s by improvements in experimental techniques, particularly those having to do with high pressure. In 1959 P. S. DeCarli and J. C. Jamieson embarked on a program to study shock metamorphism, the profound changes created in some materials by explosive shock. After quartz crystals were subjected to shock pressures of slightly more than 600 kilobars, the recovered fragments were found to have lost their crystal structure. They were amorphous—transformed to glass. The fact that the glass remained unchanged after the shock—that it did not revert back to the crystal form—was attributed to the sluggishness of the recrystallization process. A similar transformation to glass occurred when a sodium feldspar (albite: $NaAlSi_3O_8$) was similarly shocked. DeCarli and Jamieson then turned their attention to graphite; in 1961 they exposed samples of low-density graphite to shock pressures of 300 kilobars at temperatures of about 1,300°K. The result was the instantaneous creation of very small particles resembling the black diamonds (or carbonados) found in meteorites. The particles proved indeed to be diamonds.

In the same year, Lipschutz and Anders made experimental tests of several ways in which they thought that diamonds might have formed in the Canyon Diablo iron during the shock of impact. None of their experiments produced diamond; but the dramatic verification of the possibility of direct solid-state conversion of graphite to diamond by explosive shock achieved by DeCarli and Jamieson seemed, as they said, "to favor the mechanism of a direct conversion of graphite to diamond by impact shock." In their opinion, the terrestrial formation of the Canyon Diablo diamonds was by then established beyond reasonable doubt. They felt less certain, however, about the origin of diamonds in stony meteorites.

At that time (1961) diamonds had been found in only two stony meteorites, Novo Urei and Goalpara. These were both ureilites, a rare type of achondrite, and they were too small to have collided with the earth at velocities sufficiently high for the formation of diamond. They nevertheless bore indications of shock, and Lipschutz and Anders remarked that "we do know that diamonds can be formed in a meteorite by impact shock, and it is certain that the ureilites experienced at least one catastrophic impact during their pre-terrestrial history."[19] Examination of the third known ureilite, the Dyalpur meteorite, by Lipschutz in 1962 revealed that it, too, contained diamonds. "All observations to date," he concluded, "are consistent with the view that the ureilite diamonds were formed by shock during the break-up of the meteorite parent body."[20] (Discovery of diamonds in a fourth stony meteorite was reported in 1963 by two Russian investigators, in whose opinion all meteoritic diamonds had probably been formed during preterrestrial collisions of the parent bodies.[21]) In 1964 Lipschutz announced that his investigations had revealed a pronounced crystallographic orientation in diamonds from two of the ureilites, Goalpara and Novo Urei. Diamonds in the third, Dyalpur, showed no preferred orientation; but in his opinion, its textural similarities to the other two argued against a different mode of diamond formation. He concluded that "it thus appears that all meteoritic diamonds were formed by shock rather than by gravitational compression of graphite."[22]

More than a century ago, Gustav Tschermak examined the Shergotty meteorite, which fell in India in 1865. It was a basaltic chondrite, similar in texture and composition to certain gabbros (dark, coarse rocks) which occur on the earth. He discovered that it contained a glassy material which at first he thought to be a new mineral. On analysis, however, it proved to be similar in composition to plagioclase, a feldspar common in terrestrial rocks. He called this material maskelynite; and curiously enough, although it appeared to be a clear glass, it occurred in grains with rectangular outlines like those commonly seen in plagioclase crystals. In sharp contrast, the pyroxene crystals in the meteorite exhibited their normal petrographic properties.

In 1963 D. J. Milton and P. S. DeCarli obtained samples of gabbro from the Stillwater complex in Montana. This rock was chosen because it was similar in mineralogy and texture to the Shergotty meteorite. When the gabbro was exposed to shock pressures of 250 to 300 kilobars, the plagioclase within it was transformed to clear glass—maskelynite—which retained the external form of the plagioclase crystals exactly as Tschermak had observed, a century earlier, to be the case in the Shergotty meteorite. Like that in the meteorite, the pyroxene remained crystalline, evidently little affected by the shock which instantly transformed the plagioclase. Milton and DeCarli remarked that the unusual characteristics of the Shergotty meteorite could be explained if the maskelynite had formed from the plagioclase not by ordinary fusion but by solid-solid transformation, as had been reported for quartz and other materials as a result of explosive shock.

Although examples of shock metamorphism were becoming more common, the origin of the Canyon Diablo diamonds remained a matter of debate. In a paper that appeared in 1964, N. L. Carter and G. C. Kennedy seriously disagreed with the views of Lipschutz and Anders, supporting instead Urey's contention that the Canyon Diablo diamonds had formed under equilibrium conditions deep within moon-sized planetary or asteroidal bodies. A bitter controversy followed,[23] but while it was going on, additional examples of the effects of shock on various rocks and minerals continued to appear. High-pressure glasses from the Ries Basin were examined and analyzed.[24] During an investigation of impact glasses from the Ries and Aouelloul craters it was found that zircon, a mineral common in many rocks, had decomposed to silica and a fine-grained whitish material called baddeleyite ($ZnO_2$), indicating extremely high temperatures at the time of impact.[25] Synthesis of stishovite from quartz by shock wave was announced by DeCarli and Milton in 1965. Peculiar "kink-bands" were noted in the biotite of rocks shocked by nuclear explosions[26]; and similar kink-bands were found in rocks of meteorite craters. D. Stöffler, in 1966, reported investigations at the Ries Basin which indicated that evidences of high pressure and temperature gradually diminished outward from the center (i.e., from the point of impact). Similar zoning of shock metamorphism was observed by M. R. Dence around Canadian craters.[27] And eventually, perhaps because of so many dramatic demonstrations of the transformation of material by shock, the diamonds in the Canyon Diablo iron were generally conceded to have been formed by the shock of impact; and the diamond controversy died away.

It was clear that the hypervelocity impact of a large meteorite could cause instantaneous transformation of minerals within the target rocks—without melting—into an amorphous state; that is, into glass. Such glass, formed in the solid state through the action of a shock wave, is now known as diaplectic glass. Evidence of shock metamorphism in terrestrial rocks is now regarded as proof of meteoritic impact, and the shock-induced changes are known as impact metamorphism. And while study of it went on during the 1960s, the list of probable sites of meteoritic impact lengthened as new localities were discovered. Amid continuing controversy, the various criteria that had been achieved for identification of sites of meteoritic impact were called upon in attempts to determine the origin of some of the most geologically puzzling structures yet known—the Vredefort dome and the Bushveld Igneous Complex in South Africa, and Sudbury in Canada.

## CHAPTER 14

# The Bushveld Igneous Complex and the Vredefort Ring

ACCORDING TO A DESCRIPTION published in 1925 by two prominent South African geologists, the Vredefort Mountain Land, approximately 100 miles south of Pretoria, is a unique geological feature, unlike anything else known in the world. Speculating about its
origin, the South African investigators suggested that the ancient
granite which forms a plain in its center, and which had apparently
caused the upturning and overturning of rocks around it, might have
been forced upward by a column of rising magma. A few years later,
P. B. King compared the Sierra Madera dome in Texas to the Vredefort
structure and suggested a similar origin. In both cases, the suggested
explanations were not unlike Bucher's views on cryptovolcanism. The
radically different ideas of Boon and Albritton, however, interpreted
the cryptovolcanic structures listed by Bucher, and the Sierra Madera
and Vredefort domes in addition, as scars of meteoritic impact. This
suggestion was not well received by South African geologists; and, as
we have seen, it remained almost unnoticed in the United States for
more than a decade.

The Vredefort Mountain Land is only one of many remarkable
geological features in South Africa. Another, to the north of Pretoria,
is the Bushveld Igneous Complex, where a broad expanse of highly
complicated rocks has drawn world-wide attention. In fact, the Shaler
Memorial Expedition from Harvard University was organized in 1922
with the express purpose of studying the central Transvaal because,
as R. A. Daly remarked, this colossal assemblage of igneous rock was
obviously of great importance to the development of theories concerning the origin of rocks. It was not suggested until many years

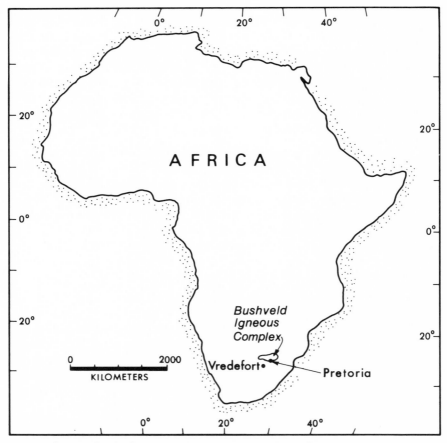

**14.1**    *Index map showing location of the Bushveld Igneous Complex.*

later that these two interesting and curious geological features might
have a common origin.

During the late 1800s unusual rocks had been reported from time
to time in the Transvaal, as well as puzzling associations of rock types
and immense accumulations of granite. Complicated outcrops and
indications of buried masses of rock could be found over an enormous
area; and the first recognition of the fact that they were related as a
whole is attributed to G. A. F. Molengraaff who, in 1898, standing at a
high point in the hills north of Pretoria, became aware of the essentially basin-shaped structure of the region.

The great assemblage of igneous rock occupying the basin became known as the Main, or Inner, Complex and proved to be more or
less pear-shaped. In some places it was covered by later (Karroo)
formations; but it was estimated to be about 300 miles long and more
than 150 miles across its widest part, and to have an area of more

than 23,000 square miles—almost equal to that of Scotland. (Later estimates suggested an area of about 26,000 square miles.) Almost completely encircling it was a belt of high ground with a consistent inward slope, interpreted as the outward physiographical expression of geological structure. And this same geological structure resulted in a landscape whose magnificence did not escape the notice of geologists attempting to understand its underlying formations. A. L. Hall, for example, who was a prominent investigator of the region during three decades, noted also its scenic wonders:

> The stupendous grandeur of the Complex has never failed to impress those who have been fortunate enough—including the members of the Shaler Memorial Expedition in 1922—to occupy any one of the numerous points of vantage available to the student of the Bushveld. Thus, on the summit of *Signal Hill,* near Magnet Heights in Sekukuniland (5400 feet, or 1646 m. above sea level) one stands on the resistant quartzitic roof-remnant of the Complex, in a geological hardly less than a physiographical sense, and commands a panorama of surpassing beauty embracing a vista of over 50 miles (80 km.) in all directions. . . .[1]

Molengraaff interpreted this complicated body of igneous rock as a huge pluton; that is, as having been formed by the subterranean

*14.2   A portion of the Bushveld Igneous Complex.*

14.3    This NASA photograph was taken from a height of 150 nautical miles and includes most
        of the Bushveld Igneous Complex, South Africa. The dark, more or less circular structure
        in the upper center of the picture is the Pilandsberg.

consolidation of immense quantities of intrusive magma. He called it
the Plutonic Series of the Bushveld. It was frequently referred to as a
laccolith, for the magma was thought to have intruded between horizontal layers of subsurface rock. In a remarkably perceptive work first
published in 1901, Molengraaff noted various tectonic effects of the
Bushveld emplacement. He observed that the sedimentary floor on
which it rested had subsided in a downward bulge, and he attributed
the centrally directed dip observed in hills encircling the Complex to
this subsidence. He noted extensive faulting and numerous dikes. He
observed the powerful metamorphic effects of the igneous intrusion
on surrounding sedimentary rocks and attributed various rock types
to gravitational differentiation of the cooling magma. (Gravitational
differentiation may occur as magma cools, when early-forming crystals sink to the bottom, thus separating the congealing magma into
layers of different mineral composition.)

    As time went on, the igneous mass of the Bushveld was often
referred to as a lopolith, a name indicating a bowl-shaped intrusive
body and also suggesting the subsidence of the floor on which it rested.
A South African geologist, E. H. L. Schwarz, noted in 1909 that extensive tracts of once-melted material are difficult to explain as having
come from the interior of the earth. He suggested that "enormous
deluges of igneous material" such as the Bushveld Complex might be
the result of collisions with large meteorites. This suggestion seems to

have passed unnoticed. H. Kynaston, in 1910, compared the Bushveld to the great igneous sheet of Sudbury, Canada, which was also basin-shaped. But as Daly and Molengraaff remarked in 1924, accumulated evidence showed "this vast body to be neither a simple laccolith nor a simple lopolith, but a group of igneous units which together may best, for the present at least, be called a 'complex'."[2] A. L. Hall pointed out in 1932 that because some of the great sheets and flows of rock in the region had undoubtedly been extruded at the surface, the terms "Bushveld Plutonic Complex," "Bushveld Laccolith," and "Bushveld Lopolith"—all of which implied subsurface intrusion only—were not applicable. "No single word exists," he wrote, "to do simultaneous justice to sills, dykes, stocks, surface flows, etc. The igneous province taken *ensemble* is neither a laccolite nor a lopolith."[3] He noted that the widely accepted term "Bushveld Igneous Complex" had first appeared in the literature in 1905.

Mapping of the Bushveld Complex commenced with the establishment of a Geological Survey in 1902. Daly and Molengraaff remarked in 1924 that the resulting maps "illustrate eloquently not only the imposing facts of the Transvaal geology, but as well, the skill and arduous labors of their authors—Hall, Kynaston, Mellor, Humphrey, and Wagner."[4] The immense task was almost completed when the Shaler Memorial Expedition visited the scene some twenty years later, and by that time the investigators were confronted by many perplexing and controversial problems. In 1922 Hall conducted the members of the expedition on a five-week tour through Sekukuniland in the northeast Transvaal where, in his words, "nature has traced the framework of the Complex and exposed its fundamental problems on a scale not approached by any other sector of the Bushveld."[5]

The Complex was estimated to be about two billion years old, and efforts to understand its origin were complicated by apparently contradictory observations. According to a widely held opinion, emplacement of the huge body of intrusive rock had been preceded by volcanism, evidence of which lay in volcanic material below the sedimentary deposits on which the Complex rested. Emplacement had been followed by the "sill phase," when sediments now forming the floor of the Complex were invaded by sheets of magma that eventually formed layers of dark, heavy rock. Interpretation of the next phase, however, involved a conflict of opinion. In the view of some investigators, the Rooiberg formation, composed of light-colored rock called felsite, had been extruded from unknown volcanic vents (presumably hidden beneath the coarse red granite of the Complex) and spread over extensive areas. Granophyre, a rock of similar composition, was known to lie immediately beneath the felsite, and the gradual transition from granophyre to felsite provided confirmation of the similar volcanic origin of the two. Other investigators, however, considered the felsites and granophyres to be intrusive—part of the "salic branch," that is, the light-colored granitic component—of the parent Bushveld magma.

As the great mass of intrusive magma congealed, gradually forming a huge saucer-shaped mass, layers of different kinds of rock developed through gravitational differentiation, building up great thicknesses of norite and related dark, heavy rocks. (Norite is a gabbro containing an iron-magnesium silicate called hypersthene.) The designation of "lopolith" was generally applied to the saucer-shaped norite mass without arousing objections. R. A. Daly remarked in 1928, "For vastness of scale and for drastic results of differentiation in place the great basic member of the Complex seems to be without peer among the known rock bodies of the world."[6] It was thought that the lighter, quartz-bearing components of the magma eventually cooled as an upper granitic layer, forming the widespread coarse red granite of the Complex and also (in the opinion of some investigators) the felsites and granophyres; and evidence supporting this view was observed in localities where clear instances of gradual transition from red granite to granophyre and felsite occurred.

Some of the most serious differences of opinion concerning features of the Complex had to do with the emplacement of the granitic rocks. During two decades of exploration and mapping, many intrusions of red granite into the norite were recorded, and the norite was known to be cut by broad granitic dikes and plugs. The red granite was also known to form intrusions into the overlying granophyre and felsite. These observations indicated that the red granite was younger than both the norite and the granophyre-felsite layers; and that, after the emplacement of the norite, the granite had presumably welled up in a separate magmatic influx from some deep-seated source concealed beneath the Complex. The idea that there had been two magmatic intrusions—one which formed the norite, and a slightly later one which cut through it, forming dikes and plugs and the widespread layer of red granite—was accepted by many investigators. But opinions remained divided.

Daly came to the conclusion that the great igneous mass of the Bushveld was not an intrusion at all. He suggested that it had been extruded at the surface as an enormous lava flow. The physical continuity observed in various localities between the different layers indicated that they had formed from one mass; and if, he remarked, "it be assumed that the roof of the coarse granite was essentially its own chilled surface material, gradually thickened by successive overflows of the same magma, a possible explanation of the somewhat contradictory observed relations among the salic members of the Complex is at hand."[7] He attributed the compactness of the felsite to rapid cooling against the air. During the cooling process, which lasted many hundreds of years, there must have been much shifting and rearrangement between the solid and the still-liquid phases within the vast magma body. He pointed out:

Local diking of one phase by another is thus here, as so often elsewhere among large eruptive bodies of the world, no necessary indication of essential difference of age between the respective phases. In each case the diking may rather be the inevitable result of the slow crystallization of a single voluminous body of magma that was subject to strains as it solidified. [8]

In the opinion of most other investigators, however, the Bushveld Igneous Complex was intrusive; and because it was the largest known assemblage of such rocks, unusual phenomena were perhaps to be expected there. And they were indeed discovered. One was widespread thermal metamorphism of such striking intensity that in some localities sediments several thousand feet thick had been affected. In some places, recrystallization of sedimentary rocks was reported to have been so thorough as to almost obliterate the bedding planes. (Recrystallization is the formation, usually due to some combination of heat and pressure, of new mineral grains in a rock without melting, as, for instance, when limestone composed of calcite recrystallizes to marble composed of calcite. In some cases, the combinations of the mineral components of the rock may change, resulting in the formation of new minerals.) Investigators observed a great variety of metamorphic rocks in many different locations. They noted that their alteration was apparently unrelated either to the presence of granitic plugs (for it often occurred where there were no granitic outcrops) or to the numerous intrusive sills, the effects of which could be observed only at the immediate contacts (that is, where the two kinds of rock came together). It gradually became clear that the metamorphism must have been induced by heat which came from the Complex itself, and that the intensity of its effects on the invaded rocks was related to their proximity to the Complex. In some locations there was evidence of pressure-caused metamorphism as well as that caused by heat.

According to Hall, the Bushveld aureole—that is, the zone in which contact metamorphism of the country rock had taken place— "arrests the attention of anyone traversing the floor up to the margin of the Complex along any section." [9] It was found to be narrowest near Pretoria and to reach its maximum values at the extreme eastern and western ends of the Complex. And its most surprising feature was a remarkable variation in the vertical range of the affected beds, which might after all be expected to have some relation to the thickness of the intrusive rock. But this seemed not to be the case. Hall pointed out, for instance, that at Zeerust, where the norite thinned to only

**14.4**    *Aerial photograph of the Pilandsberg, an intrusive structure within the Bushveld Igneous Complex, South Africa.*

5,000 feet (about half the thickness that produced the narrow aureole at Pretoria), the contact belt reached an enormous thickness—more than 10,000 feet. "Some other factor," Hall commented, "appears to come into play here."[10]

In many localities the Bushveld rock was found to contain xenoliths; that is, fragments of older rock which had become included in the magma before it congealed. They ranged in size from very small to very large. Some were enormous. Sedimentary blocks several square miles in area were observed to be entirely surrounded by formations belonging to the Complex. W. A. Humphrey in 1909 interpreted such displaced blocks as huge fragments broken off the sedimentary floor on which the Complex rested, comparable in principle to huge xenoliths; and he attributed their position to faulting. Other such blocks were reported, of similar enormous size—large enough, in fact, to appear on the geologic maps. Later investigators agreed that they were fragments of the sedimentary floor but that, being of lower density than the intruding magma, they had floated up through it into their present positions.

Valuable deposits of tin ore were discovered in the Bushveld during the early years, and as exploration continued, other economically important ores were found. H. Kynaston, in 1910, attributed their formation to hydrothermal processes; that is, to the action of hot aqueous solutions. Regionally persistent layers of chromite (chromium ore) were found to conform with widespread (and poorly understood) "rifting," a type of layering known in these igneous rocks as pseudostratification. Platinum was discovered in the early 1920s, and the ensuing search for it and other precious metals stimulated further exploration of the whole region. Evidence of diastrophism (crustal movement) was noted by Kynaston, who reported complicated folding and overthrusting in the Waterberg district. He noted, in particular, that the edges of the great Bushveld and Waterberg basins had been pushed against one another, causing dislocation, overthrusting, and folding of the strata.[11]

There were various outcrops of alkaline rocks within the Bushveld Complex. The most spectacular, studied by S. J. Shand, was the Pilandsberg, a circular structure which rose from the almost flat Bushveld like a great mountainous island. Shand found that instead of being grouped around the center of the mountain, the highest peaks occurred near its edge. Ring structure, a prominent element in the over-all makeup of the Pilandsberg, was found to be reflected in curious ring-valleys. Shand described the intense interest shared by investigators during an excursion to these outcrops:

> If there is one branch of geology that was more strongly represented than another at the Fifteenth International Geological Congress in Pretoria, it is petrology. It is true that the petrologists did not present many papers, and that their official discussions were rather scamped, but it was on the excursions that they showed their real strength. I shall not soon forget the day when I conducted a party of fifty to Pilansberg.[12] To see them jump from the buses at every stop; to hear fifty hammers ringing at once on my favorite nepheline rocks; to see the collecting bags swelling and the buses sinking lower and lower on their springs; few sights have ever given me keener pleasure. . . .[13]

Several occurrences of rock similar to that of the main Complex were found outside the basin. A notable one was a colossal dike-like intrusion, known as the Great Dyke of Southern Rhodesia. It rose some 200 to 1,000 feet above the surface rocks on either side, varying in width from less than 3 to more than 6 miles, and was of remarkable length, running for more than 300 miles in a north-northeast direction. "In several respects," remarked Hall, "this dyke is of exceptional interest: The striking similarity of the rocks to those of the Bushveld norite, their thoroughgoing differentiation, the enormous length of the intrusion and its exceptional thickness, the straight alignment. . . ."[14]

Another puzzling region was that of the Vredefort dome, which Daly called one of the most startling earth structures ever recorded in geology. Although study of the Bushveld Igneous Complex was the main object of the Shaler Memorial Expedition, it had been decided also to study other problems which might be considered exceptionally attractive. The singularly fascinating geological problems of the Vredefort region fell into this category. Molengraaff included them in his program and found an enthusiastic collaborator in A. L. Hall.

Some seventy miles south-southwest of Johannesburg, the monotonous landscape of the Northern Orange Free State—as described by local investigators—was abruptly terminated by an imposing stretch of hills and ranges known as the Vredefort Mountain Land. Like a gigantic amphitheater, these picturesque hills curved in a band more than ten miles wide around the outer edge of an almost circular granite plain—although to the south both the granite and the encircling hills were obscured by later (Karroo) formations. Investigation revealed that layers of regional subsurface rock to a depth of about six miles were sharply upturned all around the granite, and that the encircling hills had apparently been created by differential erosion of the upturned layers. The granite plain, some twenty-five miles wide, was interpreted as the upper surface of a huge pluton, or boss. This granite, in the opinion of Hall and Molengraaff, must have been uptruded "to a level which, before it was lowered again by denudation, certainly must have been much higher than that occupied by the uppermost sediments of the Transvaal System."[15] They estimated that the total vertical displacement of granite in the central portion of the Vredefort dome was more than eight miles, and adjacent sediments were found to have undergone severe metamorphism. Perhaps the most remarkable feature of this very remarkable structure, however, was the existence in the upturned layers of great blocks of rock which had been steeply tilted and in many cases overtilted, so that they sloped inward toward the granite. The overtilting was easily recognized by inversion of the normal sequence of layers within the rocks and, for good measure, was confirmed in some places by inverted ripple marks.

Not surprisingly, this region attracted much attention. Hall and Molengraaff, whose detailed study of it was published in 1925, remarked that to anyone occupying a suitably high point within the central granite plain and studying the regular amphitheater determining his horizon,

> the grand simplicity of the design at once suggests a corresponding broad simplicity in the major cause; almost inevita-

bly one is reminded of the ring-like features displayed by the craters of the moon, or of the physiography accompanying a gigantic caldera, of the "Maare" of the Laacher See in the volcanic region of the Eifel, or even of the geological history of the Rieskessel in Bavaria. The authors do not wish to suggest, however, that such analogies necessarily possess genetic significance in the case of the Vredefort Mountain Land. [16]

The authors noted that an outstanding feature of the conclusions reached by earlier investigators was a remarkable divergence of opinion. Field work had resulted in evidence which appeared to be contradictory. In the opinion of various observers, the granite had invaded the sedimentary beds, causing both the metamorphism of adjacent rocks and the overtilting of the surrounding rock layers. However, the granite was generally accepted as being identical with the old basement granite of the region, which normally lay at a depth of at least six miles below the surface. The granite was thus much older than the sedimentary layers. It therefore could not have invaded them. What, then, could have caused the metamorphism and overtilting of the sedimentary rocks around the granite boss?

After much study and with, as they said, "due sense of the great difficulties of arriving at conclusions likely to command general assent,"[17] Hall and Molengraaff became convinced that the granite was not intrusive in the sedimentary rock. Their opinion was based on

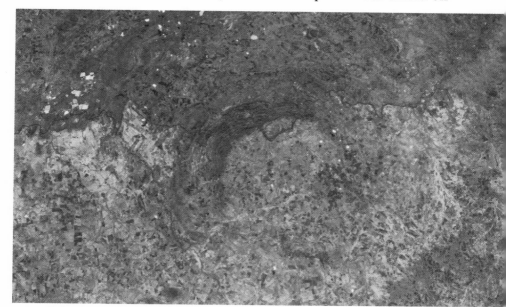

**14.5**   *The Vredefort ring, South Africa. The southern part of this ring-like structure of ancient rocks is obscured by later formations. This NASA photograph was taken from a height of 150 nautical miles.*

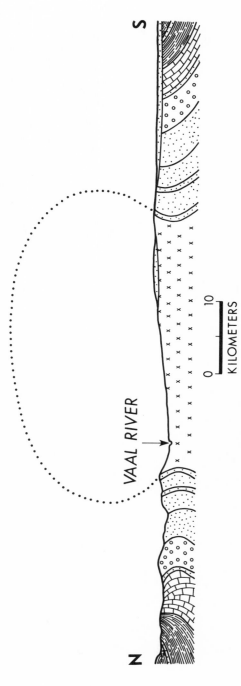

**14.6** *An ideal section across the Vredefort dome and its encircling girdle of overturned sediments, as suggested by A. L. Hall and G. A. F. Molengraaff in 1925. They pointed out that the shape of the dome, indicated by the dotted line, was entirely speculative.*

N

S

VAAL RIVER

0    10

KILOMETERS

several observations. One was the extent of the aureole, which was found to stretch outward all around the granite for the astonishing distance of at least four miles. As the granite was only twenty-five miles in diameter, an aureole of such width would "indicate a ratio between cause and effect out of all reasonable proportion and imply a stupendous source of energy; hence this line of reasoning leads to the same conclusion that the metamorphism is not due to the central granite."[18] In addition, if the metamorphism had been caused by the granite, the effects would be expected to show a progressive decrease in intensity outward in all directions from the edge of the boss. This was not the case. On the northwest side of the structure, the metamorphism was found to increase in intensity. This was attributed to the assumed subsurface location of a younger granite magma, though here too the effects were greater than seemed reasonable from this cause.

Hall and Molengraaff attributed the widespread metamorphism of the region to two components: pressure (and some heat) due to the weight of the overlying strata during the updoming of the granite; and heat (along with some pressure) supplied by other intrusions now largely concealed. Pressures capable of overturning the sedimentary beds must have been of almost incredible intensity; and as to the astonishing updoming of the central granite, the authors concluded that its initial causes were not definitely traceable. No structure with any real similarity to the Vredefort dome was known. They remarked that the closest resemblance was to be found in the Black Hills of South Dakota, and that the Rieskessel in Bavaria also showed some similarities in structure. In agreement with earlier investigators, they suggested that the updoming was probably due to some sort of tangential or centripetal pressure, along with actively rising magma at great depth below the old granite.

Various investigators suggested a possible connection between the granite of the Vredefort region and that of the Bushveld Complex. Some updoming had been observed in the Bushveld, though it was on a much smaller scale than that of the Vredefort region. Another indication of possible connection was a large number of dikes which ran from the northern edge of the Vredefort Mountain Land in a north-northwest direction into the Bushveld. Hall found it difficult to resist the conclusion that

some magmatic relationship exists between the alkaline events in the Vredefort Mountain Land and those of the Pilandsberg, based on the persistence of some of the alkaline dykes from one to the other region, and suggesting, as a reasonable speculation, a concealed but extensive source of alkaline magma.[19]

A

B

**14.7** *Pseudotachylite.* A: *Vein of pseudotachylite near Parys (Orange Free State).* B: *A boulder from one of the large veins of pseudotachylite near Parys.*

A dense, black, apparently intrusive rock was found to penetrate both the Vredefort granite and the rocks of the encircling hills. It formed dikes and networks of veins which were so numerous and widespread that Hall and Molengraaff considered them to be among the most remarkable features of the entire Vredefort Mountain Land. The veins were irregular in form and thickness and were inclined in all directions. In addition, they contained fragments of country rock ranging in size from tiny particles in the narrow veins to good-sized boulders in the dikes and larger masses. S. J. Shand, who observed the black veins and dikes in the neighborhood of Parijs (now spelled Parys), thought at first that they were intrusions of tachylite, a glass of basaltic composition. "The best exposures," he remarked in 1916, "are seen in the beds and banks of the Vaal, where the scour of the running water cleans and smooths the rock surfaces; the veins then show up jet-black and exhibit a highly polished surface, thus affording a strong contrast to the rougher grey surface of the granite."[20]

Further study indicated that the black intrusive rock was not a glass, and Shand called it pseudotachylite. In his opinion, it was different from what was known as "flinty crush-rock," a dark vein material known to occur in widely separated locations, and different also from a vein material of similar appearance in "trap-shotten gneiss," to both of which it had been compared. Flinty crush-rock had been observed in highly dislocated regions, along or near fault planes or zones of crushing, where fused portions were thought to result from heat generated by crushing. Investigation of samples of trap-shotten gneiss in veined gneiss from Salem, India, had revealed that the intrusive material was an indurated black dust containing fragments of quartz and feldspar, and that this material could be duplicated by crushing the gneiss to powder and heating it. It was thought that the veins were caused not by injection of trap (basalt), but by violent brecciation of the gneiss accompanied by heat.

Analysis of the pseudotachylite showed that its composition was such as might result, as Shand put it, from the "commingling of fragments" of the granitic rocks. Its opacity was found to be caused by a multitude of minute black specks of magnetite. Fragments of quartz and feldspar enclosed in veins of pseudotachylite were found to be much cracked; and the feldspar fragments, often minutely granulated in a manner such as might have been caused by rapid heating of the crystals, were in many cases found to be partly melted. But although the granite contained a large proportion of biotite (black mica), in no case did Shand find the biotite preserved as fragments in the vein rock. "In view of the evidence of high temperatures shown by the melted feldspar this becomes intelligible," he remarked; "biotite is decomposed by moderate heat, and its decomposition has furnished the abundant magnetite of the base of the rock."[21]

In Shand's opinion, most flinty crush-rocks were unmistakable mylonites; that is, fine-grained rocks formed by dynamic metamorphism. He felt, however, that there were weighty arguments against suggestions that pseudotachylite might be an extreme form of flinty crush rock. One such argument lay in the fact that although it contained inclusions of fragmental material, he could find no proof of origin by crushing. Another was the complete absence of dislocation and shearing in the enclosing granite, and Shand suggested that the fractures might have been caused by sudden snapping of the granite in response to sudden stress. He felt that similarly cogent arguments could be made against a possible igneous origin for the pseudotachylite. The most important of them, perhaps, had to do with the melting of the feldspar which, he remarked, "is a phenomenon for which I can recollect no parallel among igneous rocks; it seems to demand a temperature higher than one can concede to a thin tongue of magma far from its source."[22]

At the end of his painstaking investigation, the only point on which Shand felt fully satisfied was that of the intrusive relation of the pseudotachylite to the rock in which it occurred. He felt that the observed facts were better explained by regarding it not as an intrusive magma, but as a rock melt. And that left unanswered the question as to the cause of the melting. How could sufficient heat have been produced by sudden rupture of the granite without long-continued friction and shearing? As a final possibility, he noted that, in the opinion of some geologists, earthquakes are caused by deep-seated explosions; and he remarked that "there is abundant evidence that the sub-crust of these parts of South Africa has in the past contained highly explosive material. . . ." He continued:

> The form of the pseudotachylite veins indicates that the granite was shattered by a sudden gigantic impulse or series of impulses. If this impulse were of the nature of an explosion in the sub-crust, it would have as a necessary consequence the outrush of incandescent gases through all the fissures of the granite. In these circumstances fusion of the walls of the fissures might well ensue. . . . In brief, the material of the pseudotachylite seems to be melted granite. . . . The source of the heat and the mechanics of the intrusive process remain obscure.[23]

Several months after Shand presented his account of this work, new analyses of rock samples from Parijs were made by H. F. Harwood in an attempt to settle the question of the origin of the pseudotachylite. Feeling that his earlier conclusions were confirmed by Harwood's results, Shand stated that "the pseudotachylite has originated from the granite itself through melting caused (as I have shown) not by shearing but by shock, or alternatively, by gas-fluxing."[24]

Shand's investigations were carried out using granitic rocks from the neighborhood of Parijs. Hall and Molengraaff noted, however, that the black veins and dikes were plentiful not only in the central granite, but also in the rocks of the encircling hills—in fact, throughout the entire Vredefort Mountain Land. Although, as they said, an accurate measurement of its total bulk was obviously impossible, they suggested as a rough estimate that at least 800,000,000 cubic meters (about one fifth of a cubic mile) of pseudotachylite (or flinty crush-rock, as they preferred to call it) existed in veins and dikes of the Vredefort region. They remarked that although "Shand's conclusion that the pseudotachylite has originated from the granite itself through melting appears unavoidable,"[25] they did not feel prepared to endorse his opinion that the melting was caused by shock or gas-fluxing. In agreement with Shand, they found that careful examination of the intruded rocks revealed little or no evidence of dislocation or shearing; but they did observe evidence of severe crushing. In fact, crush phenomena were observed in both megascopic and microscopic examination of the Vredefort granite—caused, apparently, by "omnilateral pressures" which they interpreted as created by the weight of overlying sediments more than seven miles thick during the updoming of the granite. They attributed the cracking and crushing and the pulverization and melting of the affected rocks to these pressures. Because of their view of the process through which the pseudotachylite was formed, they did not share Shand's hesitation to regard it as a variety of flinty crush-rock. And they concluded that whether an upthrust of deep-seated magma or centripetal compression of the earth's crust was accepted as the origin of the perplexing Vredefort structure, "the reason why the updoming in both cases [i.e., in the Bushveld Complex and in the Vredefort Mountain Land] happened to take place where we see it and not somewhere else, and therefore its initial cause, remains unexplained."[26]

Investigations by the Shaler Memorial Expedition were made with the cooperation of numerous South African geologists. Resulting papers by Daly and Molengraaff (1924) and by Daly (1928) were described by A. L. Hall as "the most important contributions made toward our knowledge of the geology of the Bushveld Complex, and are based on the detailed field work carried out by the authors as members of the Shaler Memorial Expedition to South Africa during 1922."[27] Hall's outstanding work on the Bushveld Igneous Complex was also based to a large extent on the work of the Expedition. He remarked that the great trek through Sekukuniland, on which he conducted its members, left an indelible impression on his mind. "Had it not been for this unique experience," he wrote, "it is very unlikely that the writer would have responded to the suggestion of Daly and his friends that he should present an account of the Bushveld as a whole."[28]

It was an enormous task, and as assistant director of the Geological Survey, Hall had many other duties to attend to. Among them were preparations for the fifteenth meeting of the International Geological Congress in Pretoria in 1929, a labor extending over three years. He pointed out, however, that the delay in completing his account of the Complex had some advantages. One of them was that, when acting as leader of the extended excursion through the eastern Bushveld, he "presented much important field evidence to a phalanx of geologists that included the elite of international petrographical thought"[29]; and—evidently with the differences of opinion among South African geologists in mind—he remarked that the fact that the examination of this evidence gave rise to a good deal of divergent interpretation was a source of some consolation to him. Another advantage of the delay was that, before he had finished his account, the mapping of the Bushveld was completed. Hall's exhaustive study of the region was published in 1932, after his retirement from the Geological Survey.

Although the principal features of the Bushveld Igneous Complex and the Vredefort Mountain Land had been established by pioneer investigators, their writings were not readily available to geologists outside of South Africa. One of the purposes of the Shaler Memorial Expedition was to make these remarkable features better known. Without doubt, the works mentioned above assisted in accomplishing this purpose, by listing the findings of early investigators, by noting observations and contradictory opinions, and most of all, perhaps, by clearly delineating the continuing geological problems.

One indication that efforts to acquaint geologists with the wonders of South Africa were meeting with some success was evident in P. B. King's account of the Sierra Madera dome in Texas. As we have seen, he remarked in 1930 that the dome of the Sierra Madera invited comparison with the Vredefort dome of South Africa as described by Hall and Molengraaff in 1925. Although the Sierra Madera was much smaller, King noted that among other similarities to the Vredefort structure, it was symmetrical in plan, was encircled by steeply dipping and overturned strata, and exhibited faulting similar to that observed at Vredefort. Noting that Hall and Molengraaff considered the Vredefort dome to have been produced by centripetal pressures, along with the intrusion of subterranean magma, King remarked:

> It is possible that similar forces acted to produce the Sierra Madera, and that igneous intrusions played a part in the up-doming. No other force is known which could produce the radial thrusting displayed in the strata, and in no other way could the space evacuated by the uplifted strata be filled.[30]

But before long, J. D. Boon and C. C. Albritton, who evidently were also acquainted with the work of the Shaler Memorial Expedition, suggested the possibility of another force. They pointed out that Hall and Molengraaff, in their account of the Vredefort structure, "proposed no motivating cause for the development of the centripetal pressure and emphasized the difficulty of accounting for an almost circular dome by appealing to tangential stresses."[31] Pondering over the lack of obvious evidence for the falling of meteorites on the earth during geological antiquity, they attempted to predict what general type of structure would underlie a large meteorite crater. As already noted, they concluded that the structure to be expected would be a central dome, produced by elastic rebound after the enormous compression of the rocks during impact, with one or more ring folds surrounding it.[32] In such a structure, brecciation and pulverization of the rock would be widespread, providing evidence of the violence of the deforming forces. This description seemed to fit the Sierra Madera dome as well as the cryptovolcanic features described by Bucher. It also seemed to fit the Vredefort structure.

In the opinion of Boon and Albritton, the presence in the veins and dikes of the Vredefort region of 800,000,000 cubic meters of crushed and fragmented rock, as estimated by Hall and Molengraaff,

> sets the Vredefort area sharply apart from superficially similar structures such as the Black Hills dome. Could anything other than a violent explosion have produced this vast amount of crushed rock? Shand has suggested that the flinty crush-rock was formed as a result of shock caused by a 'gigantic impulse or series of impulses.' Is it possible that this impulse was the impact and resulting explosion of a gigantic meteorite which struck the area in pre-Carboniferous time?[33]

They were bold enough to think so. And in 1938 they listed the Sierra Madera structure (as described by P. B. King) and the Vredefort dome (and Flynn Creek, Tennessee), along with Bucher's list of cryptovolcanic structures, as "questionable meteoritic structures—ancient, and deeply eroded domical structures showing strong, localized deformation presumably produced by explosion or shock."[34]

It was not until a decade later that the ideas of Boon and Albritton received serious attention; and then, in 1947, they were recalled by Daly while he was reviewing the assorted theories that had been advanced to explain the Vredefort dome. Among them was that of E. B. Bailey (1926), who compared the Vredefort structure with regions in Scotland where he claimed that there was evidence that updoming had been caused by rising magma. T. L. Nel, in a memoir accompanying his map of the Vredefort region, noted that various problems remained unsolved; it was difficult, for instance, to see how overtilting on the observed scale could be caused by an upward-pressing magma even if the required energy were somehow supplied. He noted, however, that this and other problems would no doubt be solved by further study. He also observed a cone-in-cone structure in the Vredefort region which he attributed to a type of slickensiding (that is, a polishing and scratching of the rock surface resulting from friction on a fault plane).

In 1941 D. W. Bishopp noted difficulties in accepting the rather general assumption that the uplift of the Vredefort granite had carried with it an overlying cap of sediments "like a gigantic mushroom whose top has since been denuded off."[35] He suggested that the granitic magma broke through the overlying beds without creating an elevated outcrop. In 1945 B. D. Maree published the results of a gravimetric survey of the Vredefort structure. His work revealed that it was symmetrically arranged about an axis striking north-northwest, and he suggested that this might provide a key to the solution of the tectonic forces that produced the deformation. He attributed the migration of magma into the ring structure to isostatic inequilibrium; that is, an unequal balance of subsurface pressures. Various other opinions were offered, including the suggestion that the rocks of the Bushveld region had been affected by circulation of mineral-bearing solutions. The fact that this wide range of opinion represented the work of well-known and respected geologists was one more illustration of the complexity of the problem.

Daly pointed out that when Boon and Albritton proposed that the Vredefort structure might be the scar of a meteoritic impact, they shared a sense of "lack of collective security felt by geologits in general concerning any unpublished solution of the Vredefort problem. . . ."[36] Eventually Daly, too, formed the opinion that the most difficult problems presented by the Vredefort dome could indeed be better explained by the meteoritic hypothesis than as resulting from terrestrial forces. The enormous up-thrust granite core might have risen to its present level through some sort of isostatic adjustment after a great weight of crustal material was removed in the explosive impact. Heat created at the moment of impact might explain the extensive and puzzling thermal metamorphism, and the tilting and overturning of the rocks in the surrounding "collar" could have been caused by an explosion immediately following the impact. In spite of these carefully considered arguments, however, Daly did not feel that questions involved in the origin of the Vredefort dome were settled. "Absolute proof that the Vredefort deformation resulted from impact," he wrote, "is manifestly impossible."[37]

No great immediate interest seems to have been created by Daly's suggestion, though it was discussed at length, and critically, by Walter Bucher. (It is reported, in fact, that at an exhibition at the Metropolitan Museum of Art in 1948, Daly and Bucher got into such a violent argument over the Vredefort question that a guard requested that they leave.[38]) Some support for the meteoritic origin of the Vredefort structure was given by R. B. Baldwin in 1949; but the well-known South African geologist, Alexander du Toit, remarked in 1954 that structural analysis of the Vredefort region "points to purely tectonic agencies, and excludes the fantastic view of meteoritic impact proposed by Daly."[39]

By the late 1950s, as a result of his successful search for shatter cones at the sites of "cryptoexplosions," R. S. Dietz was convinced that they provide a definitive criterion for the identification of meteoritic impact scars, or astroblemes. His thoughts turned to Vredefort. Although he knew that most South African geologists accepted it as part

*14.8   Shatter cones from Vredefort.*

of a large regional geological structure which, in spite of its problems, could be reasonably explained in terms of conventional geological processes, Dietz wrote to several South African geologists asking about the possible existence of shatter cones in the Vredefort region. R. B. Hargraves, of Witwatersrand University, undertook a search for them. He discovered abundant and excellent examples in the overtilted rocks of the Vredefort ring, noting that they had been described by T. L. Nel in 1927 as cone-in-cone structures.

Hargraves remarked that according to Dietz (1960) intense shock waves are necessary to produce shatter cones, and that according to E. M. Shoemaker (1960) "a meteorite that strikes the ground at a speed exceeding the acoustic velocity of the rocks propagates a shock wave in the rocks."[40] (The acoustic velocity of the rock is the speed with which sound travels through it.) In the rugged northwest part of the Vredefort region, where plentiful shatter cones occurred, he observed that they were oriented so that their apices pointed to the northwest. Theoretically, the cones would be expected to point to the source of the shock. He noted that if the overtilted quartzites were restored to their original horizontal position, the shatter cones would then point to the center of the structure. And in certain rocks which had not been overtilted, the shatter cones actually did point to the center. He concluded:

> Whether the explosion was the culmination of a series of terrestrial igneous and/or tectonic events, or the almost instantaneous result of meteorite impact, remains to be proved. However, based on the empirical evidence of the shatter cones, a gigantic explosion of some sort appears to have been intimately related to the formation of the Vredefort ring structure as known at present.[41]

Dietz had no remaining doubts. He immediately announced, in agreement with the "heretical opinion of Boon, Albritton, and Daly,"[42] that in his opinion, too, the Vredefort structure had been created by meteoritic impact. And he listed various features of the Vredefort ring—as he preferred to call it—on which his opinion was based. One was the tilting and overtilting of the rocks surrounding the central granite which, he pointed out, could have been caused by radial forces spreading out from the center. Another was the bilateral symmetry of the structure. Bilateral symmetry was listed by Boon and Albritton in 1936, as we have seen, as a basic criterion for the identification of impact structures. Another feature was the astonishing width of the aureole, the altered sediments around the central granite, which was considered by Hall and Molengraaff to be out of all reasonable proportion if heat from the granite were assumed to be the cause. Such a phenomenon could be explained, Dietz declared, as resulting from a shock wave. Thermal and pressure-caused metamorphism in the surrounding rocks could also be explained as shock-induced. The pseudotachylite must have been created by shock pressures probably in excess of 100,000 kilobars, and the only known natural cause for shock pressures of such intensity was the impact of a meteorite. In Dietz's opinion, the presence of shatter cones confirmed the existence of these enormous pressures at the moment of impact.

Dietz's American colleagues greeted his announcement with lively, though cautious, interest. The reaction of South African geologists, however, was distinctly negative. "The geological relationships in space and time in the Vredefort region," remarked Arie Poldevaart in 1962, "do not favour the possibility that the Vredefort dome originated through meteor impact. . . . Shatter cones seem to be characteristic shock features, but shock may have been induced by surface or near-surface explosions, not of the meteor impact origin."[43]

In 1962 Dietz went a step further. He compared the association of the Vredefort ring and the Bushveld Igneous Complex with that of the Ries and Steinheim basins in Germany; and he suggested that the African structures, like the German ones, might have been formed by a twin impact event. A lesser asteroid, in his opinion, might have formed the Vredefort astrobleme, while a greater one pierced the crust in the Bushveld region with the result that a huge pool of magma welled up into the explosion crater, causing intense thermal metamorphism. Pseudovolcanic events, due to the cooling process, then caused formation of a roof of welded tuff (the Rooiberg felsites), and gravitative differentiation of the cooling magma formed the layers of rock to be seen today. Some lopoliths, Dietz concluded, are impact-generated and may be similar to lunar maria.[44] And he turned his attention to Sudbury, in Canada, a region that had often been compared to the Bushveld.

A few years later W. I. Manton of the University of Witwatersrand published a study of shatter cones in the Vredefort region confirming Hargrave's observation that, when the beds were restored to their original horizontal position, the axes of the cones pointed to the center of the ring. He noted that this spherical symmetry might have been caused by a shock wave radiating from an explosion; and he remarked that although the similarities of the Vredefort ring to cryptovolcanic structures "suggest an extraterrestrial origin . . . it seems to be closely interwoven in space and time with other events in the geological evolution of the subcontinent."[45] Evidence that might provide the correct explanation could, he suggested, be derived from the age of the shatter cones relative to the formation of the Vredefort structure — information which so far "has proved tantalizingly elusive." Manton's evident willingness to consider an extraterrestrial origin for the Vredefort dome was criticized by some of his South African colleagues. Commenting on his paper, B. B. Brock mentioned several points which in his opinion vitiated the impact hypothesis. "To call upon astronomy to solve our tectonic problems," he concluded with dry humor, "would be to minimize the usefulness of geology."[46] And controversy continued.

The attention of many investigators was caught by the various speculations about the origin of the Vredefort structure. After N. L. Carter's discovery in 1965 that samples of deformed quartz from the Vredefort region were similar to experimentally deformed quartz rather than to quartz from "ordinary" metamorphic rocks, a meteoritic origin for the Vredefort ring was accepted by many North American geologists. Additional evidence appeared in 1971 when H. G. Wilshire reported finding various kinds of shock deformation—microfracturing, deformation lamellae, kink-bands—in mineral grains taken from the Vredefort pseudotachylite. This clearly indicated, in his opinion, that the Vredefort pseudotachylite was a product of shock, probably caused by meteoritic impact. And in 1978 a South African investigator, J. E. J. Martini, discovered both coesite and stishovite in pseudotachylite veins of the "collar" of the Vredefort structure.

From time to time some investigators had suggested that coesite might be formed by temperatures and pressures within the earth, and in 1977 the discovery of coesite in a South African kimberlite was announced by J. R. Smyth and C. J. Hatton. They regarded the presence of coesite in kimberlite—a non-shock environment—as indicating rapid cooling during eruption, and they noted certain differences between the diatreme coesite and that found at impact sites. They remarked that "although coesite should no longer be considered as absolutely diagnostic of impact processes, the large grain size and primary habit of coesite from static pressure environments should easily distinguish it from impact-generated coesite."[47] Coesite thus remains an important indicator of impact, and the Vredefort ring is now widely acknowledged to be the scar created by the collision of the earth with an enormous meteorite.

The Bushveld Complex, however, with all its apparent contradictions, still presented many unsolved problems. Warren Hamilton in 1970 recalled Dietz's suggestion that it, too, might be the result of meteoritic impact. Many investigators had observed areas of updoming in the Bushveld, and Hamilton noted that a recent gravity survey indicated that, rather than being one great basin-shaped mass, the Complex was made up of three overlapping basins. In 1959 C. A. Cousins had suggested that some of the great blocks of rock apparently torn from the sedimentary floor of the Complex and interpreted by earlier investigators as huge xenoliths might instead be updomed parts of the floor itself. These observations indicated, in Hamilton's opinion, that "the Bushveld Complex resulted from the simultaneous

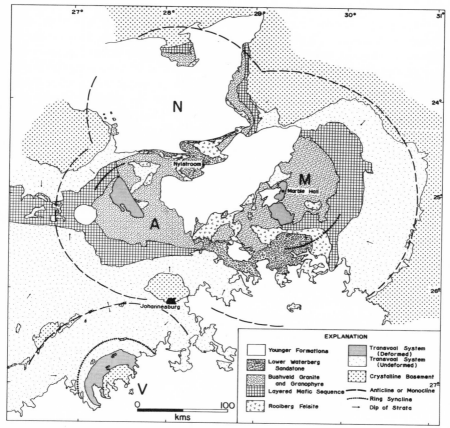

**14.9** *Generalized geologic map of the Bushveld-Vredefort complex, South Africa.*
*V = Vredefort ring; A = Assen ring; M = Marble Hall ring; N = Northern ring.*

impacts of three large masses, which produce the three overlapping circular basins and irregular central uplifts."[48] Such impacts would have produced enormous quantities of magma by fusion of crustal and mantle rocks, or by the welling up of magma through the explosively heated and thinned crust. As it had been shown that the Vredefort ring and the Bushveld Igneous Complex had approximately the same age of two billion years, Hamilton suggested that four simultaneous impacts occurred: one impact created the Vredefort structure, and the other three simultaneous impacts caused the magmatic events which created the Bushveld Complex. The rocks of the Bushveld, he remarked, should therefore be searched for evidence of shock metamorphism. He added, however, that in rock samples he had examined he recognized severe thermal metamorphism, but no shock features. This indicated that "if shock-metamorphism was present initially, it has been largely obliterated by thermal recrystallization."[49]

Further evidence for impact origin of the Bushveld Complex was advanced by Rodney C. Rhodes in 1975. He interpreted the Bushveld Complex and the Vredefort structure, in agreement with Hamilton, as the result of four simultaneous impacts. Three of them triggered extensive magmatic events which formed three separate arcuate rock bodies—which he called the Assen ring, the Marble Hall ring, and the Northern ring—that cropped out around the margin of the Complex. The fourth created the Vredefort ring. The Bushveld-Vredefort structure, in his opinion, "represents the largest hypervelocity impact feature known on earth."[50] Although no unequivocal evidence of hypervelocity impact such as shatter cones, coesite, or shock lamellae had been found in the Bushveld Complex (in spite of a careful search conducted by B. M. French and R. B. Hargraves in 1971), he pointed out that the general structure of the Bushveld-Vredefort Complex was compatible with structural patterns of other large impact features. In association with M. D. Du Plessis, Rhodes showed that the Rooiberg felsite, which was generally considered crucially important in any interpretation of the Bushveld Complex, consisted of two units. He interpreted the lower felsite unit as a layer of impact melt, produced by shock, and the quartzite blocks which it contained as almost completely melted impact breccia. It was similar, in his opinion, to layers of fused rock observed by M. R. Dence in several Canadian craters, about which Dence remarked in 1964: "Only the less altered inclusions would survive as the mass cooled in the manner of a thick lava sheet, with thermal reactions generally overwhelming the delicate products of pure shock metamorphism."[51]

During 1972 and 1973 Rhodes devoted nine months to field work in the Bushveld. The paper referred to above was a preliminary report of this work, the first of several papers in which he intended to give detailed evidence for an impact origin of the Bushveld Complex. These papers were never published, for Rhodes's work was cut short by his accidental death in 1975.

Many problems of the Bushveld Complex still await solution.

## CHAPTER 15

# Sudbury: A Geologic Puzzle

IN 1883, ORE WAS DISCOVERED near Sudbury, Ontario (Canada). Since that time several billions of dollars worth of nickel, copper, iron, and other heavy metals have been produced by mines in that region. Geological studies were undertaken in order to understand the structure and emplacement of the ore bodies, but although the investigators were competent geologists, there was much disagreement among them. They disagreed not only about the interpretation of what they saw; they disagreed about what actually had been seen. Evidence at one location contradicted evidence observed somewhere else, and attempts to understand the ore bodies became a tangle of confusion and frustration. It was not until some understanding of the effects of shock on various rocks and minerals was achieved that it was realized —with astonishment—that evidence of hypervelocity impact existed at Sudbury.

In 1882 a railway was constructed through the rugged and inhospitable Sudbury region. Thirty years earlier, a great disturbance of the compass had been observed by a land surveyor at a place in the Sudbury district which eventually became the large Creighton mine. Rock samples from the location were found to contain copper and nickel; but there was no perception of great mineral wealth until the discovery, after the advent of the railway, of large deposits of an important copper ore, chalcopyrite, a copper-iron sulfide. A curious story about the copper ore, outlined below, was related by S. J. Ritchie, of Akron, Ohio, and appeared in a 1905 report of the Ontario Bureau of Mines by A. P. Coleman.

In 1876 S. J. Ritchie (who later became one of the pioneer developers of the Sudbury nickel field) was persuaded by a somewhat eccen-

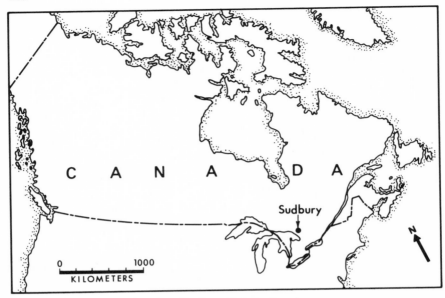

**15.1**   *Index map showing location of Sudbury, Ontario, Canada.*

tric acquaintance named John Gamgee to assist him in obtaining
government appropriations to develop a cure for yellow fever, which
was epidemic in cities of the lower Mississippi and the Gulf States. It
was thought that the germs of yellow fever were killed by frost. Gamgee
proposed to build a ship on which low temperatures would be maintained. Yellow fever patients would be taken on board and cured by the
frosty atmosphere.

Plans were drawn up, and Gamgee convinced the United States
Senate Committee on Epidemic Diseases that the enterprise was
feasible. Methods of refrigeration were not well developed at that time,
but he designed an engine that would use liquid ammonia to produce
the required temperatures. Senators and army engineers assembled to
see the engine in operation—but a snag developed because ordinary
cast iron would not contain the pressure generated by the ammonia
gas. There followed a series of experimental attempts to create alloys to
strengthen the iron and to overcome its porosity. After several weeks
of failure, Gamgee suddenly thought of the meteorites at the Smithsonian Institution:

> We have no metallic iron on earth produced by nature in that
> form, and these meteors have all fallen from the skies, or have
> come from some other world. They nearly all contain nickel,
> and are the closest grained metal we have. Tomorrow we will
> . . . try this metal as an alloy with iron, and see if we can imitate

nature in duplicating the meteorite, as we are trying to imitate nature in the production of artificial cold for the yellow fever patients.[1]

Nickel was procured, and eventually some seventy-two separate pieces of nickel-iron alloy were carefully marked to indicate that the nickel content increased from one piece to the next by 0.5 percent up to a maximum of 36 percent. The alloy containing 8 percent nickel proved to be so hard that neither file nor cold chisel had any effect upon it. According to Ritchie, Gamgee threw up his hands and shouted, "Eureka! I have found at last an alloy strong enough and hard enough to resist anything and close enough in texture to resist the escape of any form of gas!" But, Ritchie continued, "Gamgee, like most abnormally developed geniuses, had no place in his makeup for such humdrum routine efforts as financial operations . . . he failed to agree with the Senate Committee on the costs and management of the proposed ship, and thus failed to get the $250,000 appropriation."[2]

A few years later, Ritchie had an interest in construction of a railway to serve iron fields in Ontario; when the iron was found to contain too much sulfur to be marketable, it became necessary to search for other mineral deposits to support the railway. Ore specimens at the Geological Museum in Ottawa were examined, and samples taken near Sudbury were found to be very rich in copper. This led to Ritchie's involvement in the purchase of a number of Sudbury locations.

In 1886, when the products of ore from a Sudbury mine were being analyzed, the chemist found an unfamiliar metal, which eventually proved to be nickel. Ritchie remarked,

> The discovery of this nickel in these ores . . . was unexpected news. . . . We had no suspicion that they were anything but copper ores. This discovery changed the whole situation. . . . As the world's annual consumption of nickel was then only about 1,000 tons, the question was what was to be done with all the nickel which these deposits could produce. I at once recalled . . . John Gamgee in the Navy Yard at Washington . . . and it occurred to me that nickel could be used with success in the manufacture of guns and for many other purposes as an alloy with iron and steel.[3]

Ritchie wrote to the famous gunmaker, Krupp, in Essen, Germany, telling him of Gamgee's experiments, but in reply he was told that there was not enough nickel in the world to warrant experiments which pointed to increased use of it. He visited many of the great iron and steel works of Europe in order to find out whatever might be learned about uses of nickel alloys with steel, and he found that a common problem was the expense and difficulty of obtaining nickel. In addition, Ritchie drew the matter to the attention of General Tracy, Secretary of the United States Navy, with the eventual result that ordinary steel plates and nickel steel plates were tested at the government proving grounds at Annapolis, Maryland. The superiority of nickel steel was demonstrated so decisively that, at the request of General Tracy, Congress voted an appropriation of $1,000,000 to purchase nickel matte at Sudbury, to be used in the manufacture of nickel steel armor plate for the United States Navy. After that, the future economic importance of Sudbury was no longer in doubt.

All this pointed to a bright future for the Sudbury mining district and gave encouragement to prospecting and development as well as to geological explorations. The copper ore (chalcopyrite) was usually found in association with pentlandite (a nickel-iron sulfide) and with large quantities of nickeliferous pyrrhotite (an iron sulfide). The pyrrhotite was at first considered worthless, but after its value as a nickel ore became apparent, a fever of prospecting revealed scores of nickel deposits. In fact, as T. L. Walker pointed out, it was easier in the early 1890s to discover nickel properties than it was to dispose of them at remunerative prices. Things changed rapidly, however, as new demands for nickel steel created an unprecedented demand for nickel. In the decades since these discoveries were made, large quantities of nickel, copper, iron, platinum, and various other metals have been produced by Sudbury mines. Many geologists visited the district; and as a result of geological exploration and mapping, most of its surface features were fairly accurately known by the early years of the century. And by that time, as A. E. Barlow remarked in 1906, the literature of the Sudbury district was already voluminous.

A long, narrow belt, about twenty-five miles wide, of complicated Precambrian (Huronian) rock estimated to be some two billion years old, was known to extend from the north shore of Lake Huron to Lake Wanapitae and beyond. The Sudbury district was situated within this belt of rock, forming what was at first described as a north-east trending trough. Two intrusions of a coarse-textured igneous rock, for which the prospectors had several names—trap, greenstone, diorite— and which had apparently been in a molten state when it invaded the older rocks, formed the sides of the trough. Most of the earlier-discovered, heavier deposits of nickeliferous ore were associated with these intrusive belts. T. L. Walker, in 1897, showed that on the inner sides of the trough the coarse-textured greenstone or diorite graded into a pinkish rock with a granite-like appearance, which became known as micropegmatite. He explained this gradual change as due to gravitational differentiation of the original magma, and suggested that the copper-nickel deposits were the result of the same settling process. On the inner sides of the micropegmatite, on both sides of the trough, he noted a zone of what he called highly altered volcanic breccia.

In 1905 A. P. Coleman established the fact that the sides of the trough, which had been described as facing each other like a pair of brackets, were connected at the north and south ends, forming an irregular ellipse. He interpreted its elevated edge as the rim of a great laccolith—a spoon-shaped sheet of intrusive rock, the bowl of which was filled with inward-dipping volcanic tuffs and sedimentary formations which outcropped concentrically with the elliptical rim. These later became known as the Whitewater series and included, in ascending order, the Onaping tuff, Onwatin slate, and Chelmsford sandstone. Between the Onaping tuff and the micropegmatite was a layer which he called the Trout Lake conglomerate, containing flattened pebbles and a wide variety of boulders in a matrix of schist.

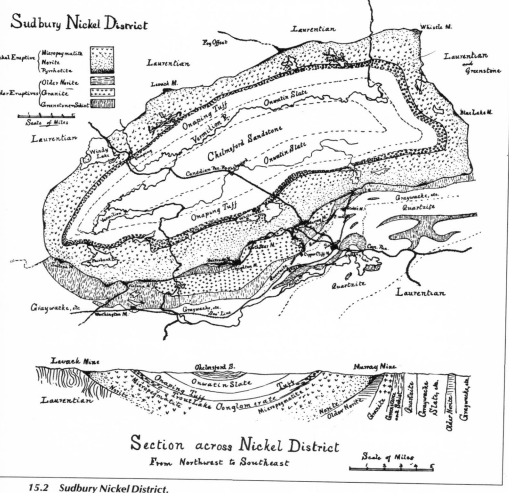

**15.2** *Sudbury Nickel District.*

"The ore bodies," Coleman remarked, "form part of the edge of a great eruptive sheet having a length of 36 miles, a breadth of 16 miles, and a thickness of a mile and a quarter. . . ."[4] His field work, begun in 1902, demonstrated that "the sheet is basin-shaped, since the lower or basic edge everywhere dips inward; and that the ore deposits are all at the lowest points of the basic edge, showing that segregation took place by the aid of gravity."[5] he noted also that unusual ores were found in some localities, and these, along with such minerals as calcite and quartz, indicated the action of water in rearranging or transporting the ore bodies. In the hills near Sudbury he observed that the gabbro was much weathered, "and wherever the adjoining sediments were found in contact with it along the edges, they are tilted up to the vertical or even overturned, so that the eruptive is clearly younger than the Huronian."[6]

Exploration revealed many puzzling aspects of the Sudbury structure. Peculiar features were revealed by microscopic examination of individual rocks, examples of which were mentioned by T. G. Bonney as early as 1888. A "silicified glass-breccia" found by Robert Bell on the northwestern side of the Huronian "trough" was described by G. H. Williams in 1891 as not merely a petrographical anomaly, but "nothing less than a breccia composed of sharply angular fragments of volcanic glass and pumice, which in spite of almost complete silicification, still preserve every detail of their original form and microlitic flow structure. . . ."[7] In addition, various indications of pressure, such as undulatory extinction in quartz grains and bending and breaking of plagioclase crystals, were observed in granite from the northwest border of the area. T. L. Walker remarked that examination of numerous specimens of ores and rocks led to contradictory conclusions as to the sequence in which they had formed. He noted also that biotite in gneiss exposed west of the Onaping Station showed a "well-developed system of partings which form a cross-hatching, probably the result of pressure."[8] And he observed in an examination of highly altered quartzites found a mile or so west of Sudbury that no sharp boundaries existed between quartz crystals within the rock, although other components appeared to be normal. "We know very little," he remarked, "as to the cause of this extreme and apparently selective metamorphism."[9] Fortunately, however, some problems were solved, rather than seeming to be intensified by microscopic examination. The type-rock of parts of the nickel-bearing intrusive—the rock variously called trap, greenstone, or diorite by the prospectors—was found to be a quartz-hypersthene-gabbro, containing small quantities of other minerals. It was, in other words, a norite. And it was in association with the norite intrusions that most of the ores were found.

Although the problems introduced by petrographical investigation of the Sudbury rocks were highly unusual and little understood, they did not at that time become a focus of attention. Instead, a heated controversy arose about another question of both scientific and economic interest: the question of the origin of the ores. Were they, as Coleman suggested, the result of gravitational differentiation within the eruptive? Or, as maintained by some later investigators, had they crystallized in a more usual manner from circulating mineralized solutions? This question was debated for decades; and it was accompanied by arguments about related problems such as the subsurface form of the igneous sheet, the age of various granitic intrusions, and whether the micropegmatite was the acidic part of the original magma or the result of a separate intrusion. It was claimed, for example, by T. C. Phemister in 1925 that a regular gradation from norite to micropegmatite, as recorded by earlier investigators, did not exist. Instead, in some localities, there was a sharp contact between the two, indicating separate intrusions. This led A. P. Coleman, E. S. Moore, and T. L. Walker (1929) to review their earlier work. After a careful investigation, they reaffirmed their opinion that the norite, micropegmatite, and ore deposits had been differentiated in place from a single body of intruded magma.

An attempt to sum up and clarify the welter of conflicting opinion was made by W. H. Collins in the mid 1930s. As he pointed out, mining had been going on for some fifty years, and opportunities for studying good exposures of the copper-nickel deposits had been greatly reduced. Some mines had ceased to operate. Many prospect openings were filled with water and mud, and weathering of their once-fresh walls had taken place rapidly due to oxidation of the sulfide minerals. He remarked that

> the geologist of today can find a wealth of information about the Sudbury ore-deposits and related rock formations in the bulky mass of writings on these subjects. Although a considerable share of these publications is theoretical and controversial, reiterative and of not much value, there is a large part, based on direct field and laboratory work, that is rich in reliable information, much of which could not be obtained now. Outstanding in this latter category are the government reports of A. P. Coleman, truly remarkable documents for the wealth and fidelity of information they contain. . . .[10]

In agreement with Walker, Coleman, and other early investigators, Collins believed that the norite and micropegmatite were formed by gravitative differentiation from a single body of homogeneous magma. He noted, nevertheless, that some conditions existed that did not harmonize with this theory. An example was the distribution of several relatively heavy mineral components of the igneous rock: apatite (a phosphate), titaniferous magnetite (an iron oxide containing titanium), and certain pyroxenes (magnesium-iron silicates)—all thought to have crystallized early in the cooling process. According to the theory of gravitational differentiation, heavier materials should be found most plentifully at the bottom of the norite. Instead, apatite was distributed in a fairly uniform manner through both the norite and the micropegmatite. The greatest concentration of titaniferous magnetite was found near the top of the norite; otherwise it, too, was disseminated uniformly through the rest of the irruptive.[11] And according to the theory of fractional crystallization, a greater downward concentration than was observed might also have been expected of the pyroxene.

Collins analyzed a great many rock samples taken in each location from the norite, the micropegmatite, and the transition zone between them. An increase in specific gravity at the transition zone, apparently due to concentration of titaniferous magnetite, was generally observed — though at the large Creighton mine Collins felt that the region of high specific gravity was too broad to be due to this cause alone. There must, he suggested, be some further reason as yet unknown. And noting some indications of subordinate differentiation within the norite, he made the following interesting observation:

> Possibly these subordinate variations within the norite layer at Sudbury are rudiments of a differentiation process that would have reached greater perfection if the irruptive had been larger and had cooled more slowly. The general likeness of the nickel irruptive to the Bushveld Complex of South Africa has been remarked by many geologists. Both are sheet-like bodies consisting of a norite layer overlain by a layer of granitic composition. The Bushveld Complex, however, is many times larger than the nickel irruptive. Its granitic layer, like the Sudbury micropegmatite, is uniform in composition, but its noritic portion is differentiated into lesser zones and layers. At the bottom is a "chilled" norite zone of fairly uniform composition. Above it is a "critical" zone composed of many layers of the utmost diversity in composition, sharply defined from one another, and arranged without any apparent regard to the law of gravitation. Above the critical zone is a "main" norite zone of fairly uniform composition, followed by a narrower zone transitional in composition towards the overlying granite. In the main zone just below the transition zone are several layers of almost pure titaniferous magnetite (Hall, 1932).

The analogy with the Sudbury nickel irruptive is highly suggestive. Both have transition zones. The zone rich in titaniferous magnetite in the nickel irruptive just below the transition zone is in the same relative position as the titaniferous iron ore layers of the Bushveld Complex. It may be that the "critical" zone at Sudbury is not well enough developed for any of the component layers to be recognizable. On the other hand, a more intensive study of the lower part of the norite might reveal similarities with the critical zone of the Bushveld Complex. . . .[12]

But in spite of this somewhat tentative comparison, the ubiquity of apatite, magnetite, and pyroxene crystals throughout the nickel irruptive remained unexplained.

Nearly all the important copper-nickel ore deposits occurred at the base of the nickel irruptive in dike-shaped downward protrusions which Coleman called "offsets." Unlike ordinary dikes, they contained a peculiar breccia made up, as Collins remarked, of "a greater variety of materials than the adjoining country rock locally affords."[13] The rock fragments, ranging in size from coarse grit to large boulders, were cemented together by what he called the "offset igneous rock." He showed that it was a quartz diorite and observed that besides its occurrence in dikes it was also found around the edges and at the bottom of the main irruptive, where it graded into the norite. "There seems to be little doubt," he remarked, "that the offsets are part of the nickel irruptive."[14] He concluded that they had formed near the end of the process of differentiation, when some of the norite magma escaped into tension cracks in the floor. Somewhat similar offsets, however, were observed to have been caused by micropegmatite entering cracks in the overlying Onaping tuff.

Various suggestions were made to account for the shape of the Sudbury irruptive. In Collins's opinion, the magma had probably been supplied by nearby vertical vents now concealed beneath the igneous rocks and had been injected between flat-lying formations where it solidified in the form of a horizontal sill. At some later time it was subjected to pressure from the south, which might have tended to produce a trough; but, as he commented, "why a pronounced basin, only 37 miles long, with fairly blunt ends, was produced can only be guessed."[15]

Collins pointed out that, with one exception, there was no dispute about structural relationships and time sequences of the Sudbury irruptive with respect to other formations in the vicinity. The exception was provided by numerous intrusions of "younger" granites, so-called to distinguish them from the ancient pre-Huronian granites of the region. Were they older or younger than the nickel irruptive? This was an important question, bearing on the origin of the ore deposits and therefore on the selection of future sites for prospecting. In some locations, the norite was penetrated by scores of granite dikes; but no dikes of norite had been observed in the granite. This indicated that the granite was the younger of the two.

If indeed the granite proved to be younger than the norite, it would follow that the ore deposits were so much younger than the nickel irruptive that they could not possibly have been derived from it. C. W. Knight, in 1916, reported hundreds of granite dikes penetrating the norite in the Creighton mine. He pointed out that since the granite was thus shown to be younger than the norite, the sulfides—the heaviest materials of the molten norite—"could not have settled to the bottom of the molten norite magma and rested on the granite footwall for the very good reason that the granite was not there when the norite was erupted."[16] And he concluded that *after* the norite was invaded by granite, the nickel and copper sulfides which form the ore body were deposited by mineralized solutions.

This evidence seemed unequivocal. "I traversed the same contact a few years later," Collins wrote, "and formed the same opinion."[17] But during a resurvey of the nickel basin in 1930, it was found that the Copper Cliff offset, a protrusion from the nickel irruptive, was intrusive in the granite; and this indicated that the granite was older than the irruptive. Because of this apparent contradiction, the whole relationship between the nickel irruptive and the younger granite was reviewed. Collins himself devoted considerable study to the apparently contradictory age relationships of the granite masses and the norite, and to the contacts between them. "The contact phenomena are also interesting in themselves," he remarked. "They are an enigma to the many able geologists who have studied them."[18]

Fine-grained or glassy igneous rock indicates rapid cooling from the original magma. A coarse texture, caused by the presence of visible crystals, indicates that the magma cooled slowly, over a relatively long period of time. Thus, fine-grained or glassy edges displayed by an intrusive rock are interpreted as the result of sudden chilling which took place as the magma penetrated older (and cooler) rock. Contacts between the granite and the nickel irruptive, however, presented a puzzling variety of features. In general, neither of them was observed to become finer near their contact, and neither had produced any apparent alteration (i.e., metamorphism) in the other. But in a few places the norite did appear to have been chilled against the granite, indicating that the granite was older.

The large accumulation of evidence indicating that the ore came from the nickel irruptive was in harmony with the idea that the granite was older than the norite; and as Collins noted, this was refuted by only one kind of evidence: the granite dikes in the norite. Perhaps, he suggested, these dikes might have been created if the younger molten norite magma had caused a marginal zone of the granite to become sufficiently plastic to flow into cracks in the norite during the cooling process. The great bulk of evidence, in any case, indicated the granite to be older than the nickel irruptive. Nevertheless, extensive observations by competent investigators indicated that the granite was both older *and* younger than the norite. "Such an absurdity," Collins remarked, "must mean that the contact phenomena on which one of the two conclusions is based have been wrongly interpreted . . . and should be reviewed."[19] Because of the remarkably similar composition of the granite masses, he did not agree with a widely accepted explanation of the problem (which claimed that there had been at least two intrusions of granite, one older and one younger than the norite) for what he called "this flat and impossible contradiction."[20]

Conspicuous blebs and spots of sulfide, which weathered into rusty-looking patches, were scattered through the offset rock and were also seen in lower parts of the norite. They were regarded by many investigators as due to chemical changes caused by circulating mineralized solutions. Collins, however, found no indication that the whole nickel irruptive had been permeated by solutions. In his opinion, the sulfide particles and spots were a primary constituent of the irruptive; and in agreement with Coleman, he maintained that the "spotted norite" in which they occurred was fresh and massive and showed little trace of the decomposing effects of solutions. "Altogether," he remarked, "there seems to be good reason to regard the spots of sulfide near the bottom of the nickel irruptive as the last of a rain of sulfides that descended through the molten magma. According to such a concept, the ore bodies would represent the fallen part of this rain."[21]

The ores were remarkably uniform throughout the large area (about 20 by 40 miles) in which they occurred. In general, they were of simple mineral composition and were found close to the nickel irruptive or its offsets. In fact, absence of the complicated minerals that are formed by mineralized solutions provided another argument in favor of gravitational differentiation. Breccia ore—rock fragments cemented together by sulfides—was found in all the large mines. The cementing sulfides were the usual mixture of pyrrhotite, chalcopyrite, and pentlandite; and the rock fragments, as Collins pointed out, did not touch each other at enough points to have been mutually supporting before being cemented together—indicating that the molten sulfides in which they were suspended probably solidified too rapidly to have had any noticeable chemical effect upon them.

Collins was well aware that his painstaking and objective study of the Sudbury district left many problems unsolved. The debate went on, and the proponents of each theory continued to cite on-the-spot observations which appeared in some cases to be flatly contradictory. In other cases, differences of opinion stemmed from differing interpretations of observed conditions. For instance, conclusions reached by A. B. Yates, chief geologist of the International Nickel Company, were quite different from those of Collins. In his opinion, the granite was unquestionably younger than the norite; and he noted in 1948 that various forms of mental gymnastics had been indulged in to jump the hurdle represented by the "significant and compelling fact" that the granite intrudes the norite. It was a paradoxical situation, for as knowledge of the Sudbury district increased, contradictory features seemed to become more numerous.

As time went on, a problem that received increasing attention was posed by the large quantities of breccia in the Sudbury region. The breccias were important because, as already mentioned, they were intimately related to the ores. Early geologists had explained the rock fragments they contained as the result of shattering caused by the intruding magma. It gradually became clear, however, that the breccia was a unique feature of the Sudbury district. No other location was known where intruding magma had caused such wholesale shattering of the country rock.

In 1942 H. W. Fairbairn and G. M. Robson observed that, in general, the Sudbury breccia consisted of patches of crushed rock which were randomly distributed, irregular in form, sometimes branching repeatedly, and ramifying in every direction through the host rock. Breccias occurred in intricate networks of veins of all sizes, and at the contacts of different kinds of rock. Although most of the rock fragments had apparently been torn from nearby formations, some had been transported for noticeable distances. In what they called "injection breccia" both the matrix and the rock fragments were foreign to the rock in which the breccia was contained. The possibility that the breccias were of volcanic origin was rejected, for no unbrecciated volcanic rocks of similar age were known at Sudbury; and they discovered only one volcanic fragment within the matrix of the breccia—a volcanic boulder probably glacially transported from elsewhere. In most cases they found a chemical and mineralogical similarity between the matrix and the rock fragments; in fact, the matrix appeared to be "comminuted material of the two rocks between which the breccia lies."[22] This finding, in their opinion, made a volcanic origin quite untenable. Also, a volcanic origin would imply that the matrices in which the fragments were embedded were molten at the time of injection; and their analyses showed that in general the matrices contained large proportions of quartz—so large, in fact, that improbably high temperatures would have to be assumed for a molten state to have existed during the formation of the breccia. They concluded that the Sudbury examples were what they called "structural breccia," which had formed during some ancient period of mountain building when the rocks were shattered by earth movements and the resulting fragments were cemented together by minerals that crystallized out of percolating groundwater. Little information was available concerning breccia of this type; and they remarked that extended field work was needed, as well as additional chemical analyses and investigation of rock fabric.

An entirely different interpretation of the Sudbury breccia was proposed in 1956 by the noted volcanologist Howel Williams. In his opinion, the Onaping tuffs (which he preferred to call tuff-breccias) had been deposited with great rapidity by glowing avalanches or nuées ardentes, similar to those which swept down from Mt. Pelée in 1902 or to those which formed the Valley of Ten Thousand Smokes in Alaska in 1912. According to Williams, glowing avalanches issued from fissures close to the inner margin of the micropegmatite, and the rapid discharge of more than 300 cubic miles of avalanche debris caused the Sudbury basin to subside, forming what he called a "deep volcano-tectonic sink." Williams recorded many observations in support of his opinion, adding still another facet to the debate about the origin of the Sudbury structure.

Different again were the ideas advocated by E. C. Speers in 1957. He saw in the Sudbury district many similarities to the Vredefort region in South Africa. He observed that the matrix of the "common" Sudbury breccia (that is, breccia in which the fragments were derived primarily from the host rock) was microscopically and chemically similar to the pseudotachylite of Vredefort and that there were many similarities in the manner in which the two occurred. He pointed out also that various investigations indicated that, at some time in the geologic past, dominant forces in the Sudbury region had created a great uprising in the underlying rock, and he noted that H. C. Cooke had observed in 1946 that the Sudbury basin lay near the crest of a major anticline. Speers gave a vivid description of what he considered to be the geologic history of the Sudbury district: mountain-building forces in the distant past created a great structural dome in the Huronian and older rocks, after which

> the central core of crystalline rocks was thrust upward as if by a punch from below, possibly by forces which resulted from magmatic action . . . and Sudbury breccias were formed in the resulting tension fractures. . . . Volcanic gases burst through the center of the dome along major rifts and thus ushered in one of the most violent periods of explosive volcanic activity ever recorded in the rocks of the earth's crust. . . . Tilting and undulation of the surface caused multitudes of tension fractures to open and close like jaws on giant crushers. Fragments which sloughed from the walls of the gaping fissures or which tumbled into the fractures from the rubble-covered surface were crushed and ground to fine sizes. . . .[23]

The great variety of opinion about the Sudbury district represented conclusions reached by competent and hard-working geologists. None of them, however, produced either evidence or arguments that were generally accepted, much less that could be considered proof of their ideas. "It may seem paradoxical, but is nevertheless true," wrote J. E. Thomson in 1957, "that the Sudbury area, despite long and intensive study, is still the center of some of the most difficult geological problems of the Precambrian shield. For over fifty years the widely accepted geological interpretation of Sudbury has been out of harmony with adjacent Precambrian areas. . . ."[24] He rejected the theory of gravitational magmatic segregation as the source of the nickel-copper deposits, noting that "this theory has already been rejected by most geologists who have written on the subject in recent years."[25] In his opinion, the general principle of a composite intrusion (that is, the formation of an intrusive complex by various components which were intruded separately) was a sounder theory; and moreover, he said, "it can be synchronized with other geological data that are being accumulated."[26]

Warren Hamilton, surveying the scene in 1961, noted the profound disagreements that were still going on concerning Sudbury, in spite of the fact that the region contained one of the most economically important and most studied rock complexes of the world. He wrote:

> It appears that "outside" geologists, and also the field geologists who did the early regional mapping of the Sudbury region, hold one set of opinions, conditioned greatly by comparisons with similar complexes in other parts of the world; whereas others, and notably many of the Sudbury geologists, are impressed by local details, and hold a conflicting set which, at the extreme, even rejects the conclusion that the Sudbury complex is a lopolith. Complicating this are the fringe opinions.[27]

According to Hamilton, the coextensiveness and evenness of the norite and granophyre (or micropegmatite, as it had more frequently been called) and the fact that the norite-granophyre, granophyre-tuff, and tuff-sediment contacts were concordant with each other and with the base of the lopolith, refuted the various arguments that they were separate intrusions. He agreed with Daly's suggestion, made a few years after his visit to South Africa, that the Bushveld Complex was *extrusive*—that it had erupted at the surface; and he noted similar indications that the Sudbury irruptive, also, was extrusive. He referred to a gravity survey carried out by A. H. Miller and M. J. S. Innes in 1955, results of which were consistent with the idea that the irruptive was a basin-shaped sheet, rather than funnel-shaped (which might indicate the upper part of a diatreme) as had been suggested by some investigators. This gave strong support to the conventional interpretation of a layered-sheet complex as suggested by Collins and Kindle in 1935, and by many others before and since. "The tidy concentricity of the map pattern and the gravity information," he remarked, "argue strongly that the lopolith is a folded sheet complex."[28]

One thing was clear: a feature of the Sudbury region that aroused no disagreement among investigators was the fact that geologically it was bewilderingly complex. "Probably only those familiar with the region are aware of the confused state of the published geological literature of Sudbury,"[29] wrote J. E. Hawley in 1962. "Even after some 70 years geologists are still debating the origin and ages of both rock formations and theories themselves."[27] After some thirty years of research, he concluded by favoring "what may be termed the older classical view of the geological history of the area, after Coleman, Collins, and Sudbury field geologists."[28]

In 1964, into this mass of confused and contradictory opinion, R. S. Dietz dropped a bombshell. Sudbury, he said, was an impact structure, an astrobleme. It was formed some 1.7 billion years ago, possibly by an iron-nickel meteorite which also contained copper, and which if traveling at about fifteen kilometers (nine miles) per second would have been almost two and a half miles in diameter. The impact formed a crater some thirty miles wide and two miles deep and triggered a magmatic event. Shock-generated heat caused melting of both local rock and meteoritic material. The removal of great quantities of rock in the excavation of the crater, along with the highly fractured condition of the crust beneath the impact site, provided easy access to the surface for magma generated in the deep crust where high temperatures prevail. The magma, upwelling from depth into the crater, formed a saucer-shaped pool—an extrusive lopolith, differentiating into layers in the crater bottom and also laying down a thick capping of welded tuffs. Water filled the resulting basin, in which the Whitewater sediments were deposited. The Grenville thrust caused pressures from the southwest, which resulted in the basin's having an oval rather than a circular shape.

This was a revolutionary idea. Tentative suggestions that the Sudbury structure might be an impact crater had indeed been made earlier, but Dietz's proposal marked the first time such a wildly unconventional interpretation had appeared in the geological literature. It was supported, however, by many new and apparently reasonable explanations for various Sudbury problems; and many geologists, after recovering their breath, felt it to be highly stimulating. Investigations to confirm or disprove the astrobleme hypothesis were immediately undertaken; but before considering them, let us first examine some of Dietz's supporting arguments.

By this time, the fact that rocks in the rims of meteorite craters are upturned and often overturned was well known. Dietz pointed out that a collar of upturned and overturned rocks, some of which had been observed by Coleman in the early days of Sudbury exploration, surrounded the southern half of the Sudbury basin. The northern half was encircled by granites. As no stratification occurs in granite, it could not be said that they were not overturned. Thus, a collar of upturned rocks existed around half of the crater and perhaps around all of it.

Speers in 1957 had described a type of Sudbury breccia as being very similar to the pseudotachylite of the Vredefort region in South Africa. As we have seen, Dietz concluded in 1961 that the pseudotachylite at Vredefort was the result of impact-shock melting. In his opinion, the Sudbury breccia, too, was a product of shock. Speers suggested that if the fissures occupied by breccia were indeed tension fractures (an opinion shared by many investigators) the direction of compression at the time of their formation must have been vertical, probably provided by an upthrust. But Dietz maintained that a vertically-directed explosive force could have come from above rather than from below. He remarked that

> two types of breccia can be anticipated: one type in which pulverized rock from the explosion crater is forced into instantaneously opened radial cracks; another type would form at focal points or along zones of weakness, where the local country rock would be crushed and injected into itself. Both of these types of breccia are found at Sudbury.[32]

Regarding the controversy concerning emplacement of the ore bodies, Dietz remarked that A. B. Yates considered magmatic segregation untenable because he had observed ore lying in sheets against the country rock and separated from the norite by a sheet of quartz diorite. This indicated that it could not have been derived from the norite. On the other hand, Hawley favored the segregation theory because of the marked dearth of hydrothermal alteration in most of the ore deposits, and also because he considered the general mineralogy to be indicative of high temperature emplacement at about 1,100°C. Dietz suggested a third possibility: ore emplacement by meteoritic impact, in which case the ore would be of cosmic rather than terrestrial origin. This, in his opinion, could simultaneously explain both Yates's and Hawley's observations: on impact, melted meteoritic material would have been splashed onto the sides of the crater, forming massive ores as well as injections into radial cracks and matrix material in breccias. Such emplacement would resemble magmatic segregation in producing anhydrous high-temperature minerals, while field relations would seem anomalous.

Some investigators, as we have seen, regarded the quartz diorite as a transition phase of the norite. Others regarded it as a separate igneous rock. Dietz suggested that it was an impact-liquified country rock which, like the melted meteoritic material, was splashed against the crater wall. He pointed out that its composition was equivalent to that of graywacke rocks (a type of sandstone) such as were thought to have constituted the Precambrian terrain in the Sudbury region. Impact emplacement would imply simultaneous origin for the ore and the quartz diorite, explaining "why their comparative age relationship had defied any clear sorting out."[33]

Considered as extrusive, the Sudbury lopolith could be regarded as a three-layered body made up of norite, micropegmatite, and Onaping tuff. Dietz suggested that the norite and micropegmatite differentiated from the magma pool within the explosion crater, forming a saucer-shaped body, while release of gas and other pseudovolcanic processes caused formation of the surface covering of Onaping tuff. Intrusions of the micropegmatite into the overlying tuff (observed by Collins and many others) had been interpreted as indicating that the tuff was emplaced before the injection of the micropegmatite; but Dietz pointed out that this relationship might only prove that tuff cooled more quickly than micropegmatite—in a matter of weeks rather than over many years. And the absence of chilled borders, which caused so much confusion in the interpretation of age relationships, might have been

**15.3** *Shatter cones in Mississagi quartzite along southern margin of the Sudbury Basin structure. Their orientation is such that, when the rocks are returned to their pre-event position, the cones point inward toward the Sudbury Basin. This suggests that the structure is an astrobleme.*

due to the elimination of great temperature differences because of heating of the country rock by impact shock.

An interesting illustration of the interlocking character of earth problems may be seen in the fact that Dietz attributed his initial interest in Sudbury to his geological studies of ocean basins. He had become convinced that no similarity exists between the oceans of the earth and the lunar maria, which were once thought to be seas. Beneath the earth's oceans a relatively narrow layer of heavy rock lies between a thin veneer of sediments and the earth's mantle. Lunar maria, in sharp contrast, appear to be large, saucer-shaped craters formed by the impact of asteroids. In his opinion, they are probably filled with impact-triggered magma. The possible existence of terrestrial structures similar to lunar maria led Dietz to investigate Sudbury as perhaps an example of an impact on earth large enough to trigger magmatism. "In this context," he remarked, "the Sudbury structure may be the terrestrial analogue of a small lunar mare."[34]

Dietz predicted that shatter cones would be found in the Sudbury region, and in May 1962, searching in the hills south of Sudbury, he fulfilled his prediction. A few months later, shatter cones were found there independently by R. B. Hargraves, who recognized them as similar to those which he had discovered in the Vredefort rocks. The search continued, and shatter cones proved to be plentiful. In 1964 R. S. Dietz and L. W. Butler examined their orientation in five locations in the steeply dipping (Mississagi) quartzite south of Sudbury. On restoring the beds to a horizontal position (thought to have been their attitude at the time of the Sudbury event) the apices of most of the shatter cones were found to point toward the Sudbury basin. It was concluded therefore that the force which shattered the rocks on the south side of the basin came from the north. And this, said Dietz and Butler, "is in accordance with expectations if the Sudbury structure is an astrobleme."[35]

By this time, interest was high among local geologists, and investigations were being carried out by J. Guy Bray and the geological staff of the International Nickel Company. They reported in 1966 that shatter cones, ranging in size from less than an inch to more than ten feet occurred in a belt some eleven miles wide around the nickel irruptive. However, Bray and his associates remarked with due caution that

> although we believe that a meteoritic origin for the Basin merits continuing consideration, and that it could be reconciled with the observed geological relationships without undue difficulty, we prefer to suspend judgment, having in mind that shatter cones have also been attributed to volcanic and tectonic activity, both of which are prominent in our interpretations of Sudbury geological history . . . and although Sudbury structure also possesses other features thought to be characteristic

of cryptoexplosion structures, much additional information must be forthcoming before the theory can be satisfactorily assessed and the shatter cones used to help unravel the geological history of the Sudbury Basin. . . .[36]

As already mentioned, study of the effects of hypervelocity impact on various rocks and minerals had established certain criteria by which shocked materials might be recognized. Among them were the presence of the high-pressure polymorphs of silica—coesite and stishovite; various deformations of quartz crystals; conversion of crystals of quartz and feldspar to glass with preservation of the original grain boundaries; kink-bands in mica; fusion and decomposition of certain opaque minerals such as zircon; and formation of different kinds of glass. These studies recalled the puzzling features recorded by the early investigators T. G. Bonney, T. L. Walker, and G. H. Williams in their microscopic examination of Sudbury rocks—features which were now recognized as shock metamorphism. Many of the specimens had been taken from what Walker described as "highly altered volcanic breccia"[37]—the Onaping tuff.

A. P. Coleman remarked in 1905 that "no rock in the Sudbury district except the nickel-bearing eruptive itself has attracted more attention than the tuff (or vitrophyre tuff) which runs as a range of hills round the outer edge of the basin. . . ."[38] During all the years of exploration, the tuff continued to attract attention. Coleman interpreted the Trout Lake conglomerate, which lay between the tuff and the micropegmatite, as a sedimentary deposit, though it graded imperceptibly into the overlying Onaping formation and was regarded by others as the base of the Onaping. Some investigators considered the Onaping tuff to be intrusive; but it was generally regarded as a volcanic deposit, perhaps the result of ash falls or of glowing avalanches. It was a black, structureless rock which had been compared to the suevite of the Ries Basin, and it contained rock fragments of various kinds, including gneisses and granites derived from the basement rocks below the irruptive. J. S. Stevenson, in 1960, observed blocks of highly altered sedimentary quartzites in this formation. In fact, in his opinion the base of the Onaping tuff was a quartzite breccia; and he observed that in some locations the matrix appeared to be an upper phase of the underlying micropegmatite, while in others the rock fragments and the matrix were identical in texture and composition.

An investigation of the Onaping tuff by B. M. French resulted in the discovery of widespread shock metamorphism.[39] He found many kinds of glass, and some of them, like those described by G. H. Williams in 1891, exhibited flow structure. He also observed glassy rims and edges around rock fragments within the tuff. Deformation was common in quartz and feldspar crystals, and some quartz crystals were found to contain microscopic fracturing. Neither coesite nor stishovite was discovered. French, however, did not feel that their absence could be considered an argument against meteorite impact; the material examined might not have been strongly enough shocked for coesite to develop, or any coesite originally formed might have been altered by subsequent metamorphism. Stishovite would not be expected to survive even mild metamorphism[40] —and at least two metamorphic events had affected the Sudbury region since the time of the impact.[41]

Speers noted strong similarities between the Sudbury breccia and the Onaping tuff. In the opinion of R. S. Dietz, as we have seen, the Sudbury breccia was a product of shock. According to French, "most if not all of the Onaping formation represents an accumulation of shocked, fractured, and partly- or completely-melted basement rocks that were excavated by the meteorite impact and subsequently redeposited. . . ."[42] In other words, the Onaping tuff was a fall-back breccia — a suevite.

As a result of these investigations, many geologists accepted a meteoritic origin for the Sudbury basin as a working hypothesis, and although some voices were raised in disagreement, Sudbury soon appeared on the list of recognized impact structures. However, Dietz's suggestion that the Sudbury ores were of cosmic origin received little support. Contamination of melt rocks by meteoritic material had, of course, been known since the discovery of nickel-iron spherules in glasses from Henbury and Wabar. M. R. Dence pointed out in 1971 that distinct nickel enrichment had been reported in the melt rocks at several impact sites, including East Clearwater Lake. But J. Guy Bray, in 1972, stated his opinion, considerably modified since 1966, as follows:

> I consider that the Sudbury Basin structure was initiated by the explosive impact of a large meteorite; I do not believe the ores came from the bolide; but I do think its impact acted as a trigger for endogenic magmatism that produced the Sudbury Nickel Irruptive, of which the ores are a part.[43]

Nevertheless, Dietz maintained that the nickel in the Sudbury ores, which were unique in their grand scale and apparently underlain by no magma chamber, must have been derived in some manner from the meteorite. He noted serious objections to the theory that the ores settled gravitationally from the irruptive as immiscible droplets—the classical "rain of sulfides." One of them was that the irruptive probably did not contain enough sulfur. Another, his basic argument, was that ores so intimately related with the sub-layer (with, that is, the norite, inclusions, and sulfides) must have been emplaced along with it. In a paper that appeared in 1972, he stated that his purpose in writing it was to update and extend the astrobleme concept of Sudbury and to attempt to further specify why an exogenic origin for the Sudbury ores remains a viable or at least a possible supposition. . . . Of course, the mechanics of impact on the scale envisioned at Sudbury remains poorly known so that almost any scheme remains possible. Many surprises await. . . .[44]

Whatever may be revealed by future work concerning the origin of the Sudbury ores, the region did indeed continue to produce surprises. Signs of a completely unexpected one were first indicated some twenty-five miles northeast of the Sudbury basin at the location of Lake Wanapitei, parts of which were known to be remarkably deep. During the winter of 1969, more than seventy gravity stations were established on the frozen surface of the lake and on islands and shoreline. The results indicated an almost circular gravity depression, centered over the circular island-free part of the lake, indicating underlying rocks of significantly lower density than any in the immediate surroundings. This discovery brought to mind the gravity anomalies associated with craters attributed to impact—Deep Bay, Saskatchewan, in particular.

The implication that Lake Wanapitei, in close proximity to Sudbury, might also be a site of meteoritic impact suggested such an improbable coincidence that there were reservations on the part of investigators about accepting the idea. But in June 1970 pebbles and boulders of mixed breccia were discovered on the lakeshore by M. R. Dence.[45] Later, rich deposits of breccia were found there by J. Guy Bray, who interpreted them as scoured from the floor of the lake by glacial action. Within the breccias, glassy igneous rocks and deformed and altered fragments were plentiful. All stages of metamorphism of quartz were observed, and the search culminated in the discovery of coesite.[46] It occurred in shock-produced glasses of various kinds, indicating that shock pressures of 425 to 500 kilobars had existed at the time of impact. Thus, in spite of what seemed to be an almost overwhelming improbability, Lake Wanapitei was identified as an impact crater—much younger, however, than the Sudbury structure. Dence and Popelar compared its relationship to the Sudbury basin with that of the lunar craters Theophilus and Theophilus B, an apparently similar association of a large ancient crater and a much smaller, more recent one.[47]

It seems highly likely that if S. J. Ritchie and John Gamgee could have had any inkling of future discoveries and developments in the Sudbury region, they would have considered themselves born several generations too soon.

## CHAPTER 16

# The Origin of Tektites and the Significance of Cratering

THE EXCITEMENT CAUSED in the 1950s by the discovery of several previously unknown "fossil" meteorite craters in Canada was shared by a rather small group of investigators. But the excitement was contagious, and the number of investigators increased rapidly as other meteorite craters were discovered. However, the recognition and investigation of meteorite craters on the earth was only one of many lines of inquiry being followed simultaneously during the mid-decades of the twentieth century. Although it is probably only in future retrospect, with a longer backward view than is yet possible, that the achievements of the period can be historically assessed, those decades of intense productivity have already been called a golden age of science. During those years, constant improvements in techniques of investigation resulted in new insights and in the solution of many old and obstinate problems. Major discoveries appeared in breathtakingly rapid succession, and developments in one field illuminated the problems of another in a dramatic illustration of the basic unity of the natural sciences.

Geology was profoundly affected by developments in other fields. Seismic and sonic methods, for example, were used in exploration of the sea floor, which revealed the world-encircling system of mid-ocean ridges and resulted in a new and revolutionary concept of the structure of the earth. The newly dawned atomic age brought methods of radioactive dating into wide use, and sites of nuclear explosions provided investigators of natural craters with previously unavailable criteria for comparison. Geologists and astronomers were drawn into closer association than ever before through their mutual interest in the composition and origin of meteorites, and in the implications suggested by the fact that both the earth and the moon had sustained heavy meteoritic bombardment in the geologic past. B. M. French pointed out in 1970 that calculations from astronomical and geological data indicated that during the past two billion years meteoritic impact in the land areas of the world had probably created about twenty craters the size of Sudbury or Vredefort, about 6,000 craters larger than three miles in diameter, and approximately 100,000 craters larger than Meteor Crater, Arizona. In his opinion "astronomical and geological data both support the present existence of perhaps several hundred large impact craters on the land areas of the earth."[1]

Discoveries continued as investigations went on. For example, natural maskelynite—glass formed from plagioclase by shock wave, similar to that found a century ago by Tschermak in the Shergotty meteorite — was found in gabbroic rock that cropped out on a small island near the center of West Clearwater Lake. Specimens of this rock were collected in 1962 by M. R. Dence and E. M. Shoemaker, who observed that it contained considerable glass. It was thought by some investigators to be volcanic, but Dence compared it with the gabbro that had been artificially shocked by Milton and DeCarli and became convinced that it was natural maskelynite.[2] A petrographic study confirmed this conclusion. Natural maskelynite was also found in the central uplift of the Manicouagan structure. As experimental evidence indicated that shock pressures of more than fifty kilobars were required to form maskelynite, its presence in these locations supported a meteoritic origin for both West Clearwater and Manicouagan lakes.

Another example was reported in 1972, when Russian investigators discovered diamonds in rocks surrounding the Popigai crater in northern Siberia.[3] This great depression, sixty miles (100 km) in diameter, had been recognized as a meteorite crater when shock-metamorphosed rocks, shatter cones, and high-pressure glasses were found in the region. Rocks previously thought to be andesite tuffs were determined instead to be impactites. The investigators pointed out that diamonds contained in the impactites were similar to meteoritic diamonds. In their opinion, they had crystallized, probably from graphite contained in gneisses of the region, under the influence of a powerful shock wave.

The characteristics of rocks associated with some fifty-two "presumed meteorite impact structures" were listed by N. M. Short and T. E. Bunch in 1968. Besides many newly dicovered locations, the list included some structures for which a meteoritic origin had long been suspected. One of them was Lonar Lake, India, where investigation was undertaken in 1964 by E. C. Lafond and R. S. Dietz. Its meteoritic origin was confirmed in 1973 with the discovery of impact glasses and other shocked material at the location, and of another smaller crater nearby, rock samples from which also showed evidence of shock. Another example was the Köfels crater, a curious widening of a valley in the Tyrolian Alps, for which a meteoritic origin was suggested in 1936 by F. E. Suess. Fused rocks from the region were examined in 1969. They indicated brief exposure to extremely high temperatures, suggesting the possibility of meteoritic impact.[4]

Nevertheless, although the lists of known sites of meteoritic impact were lengthening, recognition of them was still not a simple matter. Attempted confirmation sometimes achieved the opposite results. In the opinion of various investigators, for example, the Pretoria Salt Pan in South Africa was probably a meteorite crater; its circular shape, raised rim, and apparent absence of volcanic materials made an impact origin seem highly likely. It was felt that conclusive evidence would eventually be obtained by drilling through the lake beds on the crater floor.[5] Contrary to expectations, no positive indications supporting an impact origin were found during investigations in 1973.[6] The conclusion that the Pretoria Salt Pan was not an impact crater was supported when drilling revealed that soft saline sediments existed to a depth of more than 500 feet below the floor of the structure.

Over many years, various possible origins including that of meteoritic impact were suggested for the Carolina Bays, a number of elliptical depressions on the eastern seaboard of the United States. While these peculiar "bays" are still not well understood, the consensus of opinion at present is that they are not meteoritic.[7]

Negative results were also reached by R. S. Dietz and J. P. Barringer in 1973 in a search for evidence of impact in the region of the Hudson Bay arc. They found no shatter cones, no suevite or unusual melt rocks, no radial faults or fractures, and no metamorphic effects. They pointed out that these negative results did not disprove an impact origin for the arc, but they felt that such an origin appeared unlikely.

Another structure, located in Kansas, showed up on aerial photographs as an almost perfect circle and had been included in published lists of the world's possible meteorite craters as the Winkler crater. Investigation by D. G. Brookins in 1969 failed to yield evidence of meteoritic impact. A magnetic anomaly within the circular feature was found to be similar to those associated with several nearby kimberlite pipes, and drilling indicated the presence of kimberlite minerals within a pipelike structure. Brookins regarded the Winkler crater as probably formed by processes connected with kimberlite emplacement and recommended its removal from lists of the world's possible meteorite craters.

These examples and many others emphasized the necessity for individual examination of every structure of suspected meteoritic origin.

Closely related to the recognition of terrestrial meteorite craters was the problem presented by tektites—curiously shaped bits and pieces of natural glass, green, brown, or black, found in patches (strewn fields) in only a few regions of the world. They occurred in a great range of small sizes, averaging about that of a walnut, and various suggestions were made to explain their origin. An early idea, originating with Charles Darwin, was that they were volcanic. During the voyage of the *Beagle,* Darwin visited a number of volcanic islands, and on Ascension Island he reported finding numerous volcanic bombs. The peculiar shapes of some of them he attributed to their having been projected through the air with a "rapid rotatory motion" while partially melted. He pointed out that several other travelers had found little balls and spindles and "flattened ovals" with a "singular artificial-like appearance,"[8] objects which he thought to be composed of obsidian (volcanic glass). These, too, had probably been shot into the air during volcanic eruptions—though in some cases great numbers of them had been found in locations far from any volcano. It was later shown that the glass of which they were composed was not volcanic.

Tektites found in central Europe were at one time vaguely attributed to the activities of some ancient glass factory. Another theory suggested that they were formed by fusion of atmospheric dust by lightning. In the opinion of some investigators they had been shot from active volcanoes on the moon. For many years, however, the most prevalent view was that they were small meteoritic fragments of a strange type, although they showed no similarity to any known meteorites. The word "tektite" is derived from the Greek word *tektos,* meaning melted.

L. J. Spencer pointed out in 1933 that tektites were composed of silica glass containing small amounts of impurities. He remarked that pieces ("bombs") of silica glass from both Henbury and Wabar were very similar to tektites, and that some of them could be matched exactly in form and appearance with Darwin glass from Tasmania—the glass caused F. C. Leonard to include a suspected but undiscovered "Darwin Crater" (Mount Darwin) in his list of terrestrial meteorite craters (see Chapter 8). In Spencer's opinion, as we have seen, the minute spheres of nickel-iron which he observed in silica glass from Wabar and Henbury represented "a rain or fine drizzle of condensed iron and nickel from the vaporization of the meteorite."[9] It is interesting to note that he found similar metallic spheres "abundantly present in some specimens of Darwin glass from Tasmania"; and that in microsections of tektites from Australia and Indochina "a few black spots of similar dimensions and showing a metallic luster by reflected light can usually be found if patiently searched for. . . ."[10] He remarked in 1933 that

> the bombs of silica-glass collected by Mr. Philby at Wabar suggest there must have been a pool of molten silica in the desert sand and that this material was shot out from the craters through an atmosphere of silica, iron, and nickel

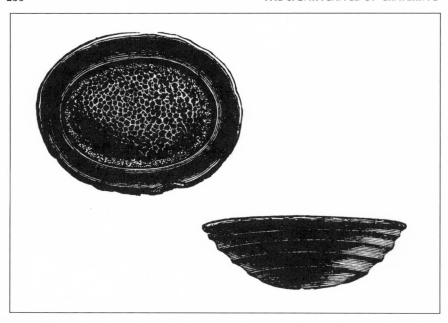

**16.1**   *Tektites from Australia thought by Darwin to be volcanic bombs. The upper figure gives a front view; the lower a side view of the same object.*

vapours. . . . I have no hesitation in concluding that tektites are not meteoric though they are connected with the fall of large meteoric masses, but that they have resulted from the fusion of terrestrial rocks, especially in sandy deserts, by the heat so developed. [11]

And he noted that because the craters of the Island of Oesel in Estonia were in dolomite, and the rocks in the region of the Tunguska fall were basaltic, in neither of these places could silica glass (and, therefore, tektites) have been formed. In Spencer's opinion, the characteristics of Libyan desert glass allied it to tektites.

In 1939 V. E. Barnes observed lechatelierite particles within tektites, indicating that they had been subjected to intense heat. But the curious form of many tektites indicated aerodynamic sculpturing, apparently during a swift flight through the air. This supported the idea that they were meteoritic, perhaps fragments of a disrupted planet. H. H. Nininger, in 1943, suggested that they were created when fused rock was splashed off the surface of the moon as a result of meteoritic impact. But the very localized regions, or strewn fields, in which tektites occurred presented the theories of cosmic origin with a serious difficulty. H. C. Urey remarked that he could find "no reasonably probable way by which tektites can come from the moon and arrive at the earth with the observed distribution of the Asian-Australian field."[12] He suggested in 1957 that they might be the result of a collision of a comet with the earth.

Tektites interested astronomers as well as geologists, and a full-blown controversy developed around the question of their origin. Hundreds of papers were written about them. In 1958 G. R. Tilton remarked that many investigators had emphasized the fact that the bulk composition of tektites was similar to argillaceous (clayey) sediments

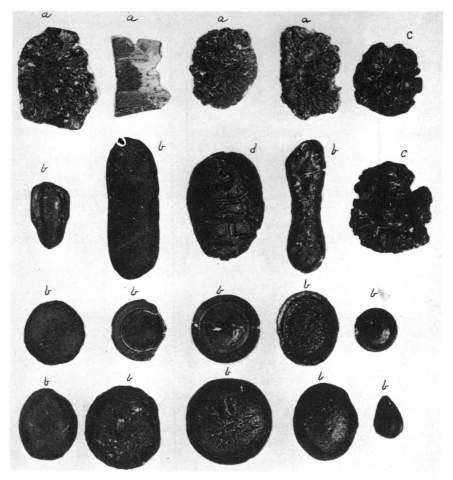

**16.2**   *Representative samples of tektites from various localities. Top row shows some sculptures and fluted examples (possibly further etched out by natural corrosion); elongate forms appear in the second row; and tear-drop and button-shaped tektites are placed in the third row. Types: a. Moldavites (from Czechoslovakia); b. Australites (from Australia); c. Bediasites (from Texas); d. Billitonites (from the Isle of Billiton). Scale: tektite in bottom center is 3 cm in diameter.*

on the earth. S. J. Rinehart pointed out that, if they originated when fused material was splashed from a meteorite crater at the moment of impact, the composition of the fused material must be that of the local soil and rock, and would probably also contain some meteoritic material.[13] And in 1962, metallic spherules containing both nickel and iron (recalling the "black spots" observed by Spencer some thirty years earlier) were found in tektites from the Philippines by E. C. T. Chao and his associates. Although this evidence strongly suggested that tektites originated during meteoritic impact, a study of their peculiar aerodynamic features caused these investigators to favor an extraterrestrial location for the impact—perhaps on the surface of the moon.

Most tektites occurred in "splash" or aerodynamic forms such as spheres, ellipsoids, dumb-bells, teardrops, and so on; but a layered variety known as the Muong Nong type, found in the general region of Thailand, occurred as chips and larger chunks. V. E. Barnes suggested that they were formed during meteoritic impact by the accumulation of melts of different viscosities in low spots on the earth's surface at the impact location. Barnes considered both the Libyan and the Darwin glasses to be tektites, and this opinion received some support from other investigators.[14] S. R. Taylor and M. Solomon, for instance, showed that the composition of Darwin glass indicated that it was not formed by terrestrial igneous processes, and they classed it as an impactite, in spite of failure to locate an impact crater in the vicinity. In 1965 coesite was discovered in Muong Nong type tektites by L. S. Walter, indicating the existence of shock pressures at the time of their formation.

It was suggested in 1963 that certain Czechoslovakian tektites might have originated in the explosion that produced the Ries crater.[15] In 1964 it was observed that Ivory Coast tektites were the same age as glass from the Bosumtwi crater.[16] The simultaneous age of certain tektites and natural glasses associated with sites of meteoritic impact was noted by several investigators.[17] These observations led Henry Faul to declare categorically in 1966 that tektites are terrestrial and that the controversy over a terrestrial versus a cosmic origin for them was nearing its end. Indeed, analysis of lunar samples in 1969 and 1970 indicated the absence on the surface of the moon of highly siliceous rocks from which tektites might have formed.[18] A terrestrial origin for tektites is now widely, though not universally, accepted; and their aerodynamic sculpturing is attributed to their extremely rapid flight through the atmosphere after being splashed from the impact site. An extraterrestrial origin, however, is still favored by a few investigators, including J. A. O'Keefe and D. R. Chapman, both of whom have devoted extensive study to tektites.

In continuing work, a possible impact crater in Tasmania, associated with Darwin glass, was reported in 1972. It was investigated in 1978 and 1979 by R. F. Fudali and R. J. Ford, who concluded that the "Darwin Crater is most likely an impact crater of rather unusual configuration, that has ejected more impact-generated melt, by a large factor, than any other known terrestrial meteorite crater of comparable size."[19] Elgygytgyn, an enormous relatively recent crater more than ten miles (18 km) in diameter discovered in northern Siberia, was suggested by R. S. Dietz as the meteorite impact site from which the tektites of the Australasian strewn field were splashed.[20] However, fission-track dating later revealed an age discrepancy between the Australasian tektites and the Elgygytgyn crater.[21]

In the late 1960s the greatest influence on crater investigation came from achievements in the space program, which itself was the result of developments in physics and of materials resistant to high pressures and temperatures—and of a wealth of newly developed technical capabilities. It brought the age-old dream of travel to outer space into the realm not only of possibility but of definite planning; and the decision to put men on the moon made it essential to determine possible landing sites and therefore to find out all that could be discovered about lunar geology and surface conditions.

Excellent telescopic photographs of the moon had already been taken from observatories on the earth, but photographs taken from the orbiting spacecraft revealed the scarred lunar surface in enormously greater detail. In 1959 a Russian spaceship achieved the first image of the moon's far side "and circumnavigated our satellite," as Zdenek Kopal remarked, "—a feat accomplished 437 years (a mere instant in cosmic history) after the first circumnavigation of our own planet by Magellan, which confirmed the spherical shape of the earth."[22] Less than a decade later, cameras of the Lunar Orbiters photographed the entire surface of the moon, its far side as well as its familiar face, resulting in a practically complete photographic coverage of the lunar surface.

Much information was obtained by the various spacecraft, but many questions awaited on-the-spot observation. Was there a possible source of water on the moon? Was there any evidence of the past or present existence of life? Were the thousands of craters, ranging from immense circular structures hundreds of miles across to those a foot or two in diameter, due mainly to impact, or was volcanism also a factor? And there were some troubling questions, too. For example, was the lunar surface capable of sustaining the weight of landing equipment, or did it conceal treacherous depths of unconsolidated dust?

A supreme human adventure—the first landing of human beings on the moon—took place in July 1969 on a surface that proved to be adequately firm. On the return trip the Apollo 11 spaceship carried a precious cargo of rocks and soil from the lunar surface. Each of the five later Apollo missions, landing at carefully chosen sites in different types of terrain, returned with samples of lunar material. These samples were distributed to laboratories where they were examined and analyzed by a multitude of scientists.

On-the-spot observation revealed no source of water on the moon, and later examination of lunar materials disclosed no water-containing rocks. To the general disappointment of biologists, no evidence of life was found. Craters were the dominant topographical form — overlapping, superimposed, covering all surfaces. Some rocks were even pitted with microcraters, invisible to the unaided eye, which indicated that the impacting meteoritic particles were sometimes of microscopic dimensions. Of all the assorted predictions concerning the lunar surface, that made in 1962 by E. M. Shoemaker seems to have come closest to observed conditions: "The lunar surface is locally built up of an imbricate and complexly overlapping set of layers composed mainly of material forming rims of craters, material underlying the crater walls and floors, and material that occupies the maria."[23]

Volcanic events had unquestionably taken place on the moon, for volcanic material was found in the lunar samples; and the lunar maria, as long suspected, were shown to be basaltic flows. But there was no doubt that the great preponderance of lunar craters were of impact origin. On the dry, still surface of the moon there were many large craters of enormous age, with outlines softened and blurred by superimposed later craters and by debris thrown out during nearby explosive impacts. Observation confirmed the evidence of telescopic photographs which showed that the lunar mountains (the terrae) were peppered with craters, displaying far more of them per unit of surface than were to be found on the maria—and thus indicating that the maria were younger than the terrae. It was suggested that, if the moon had formed by the accretion of asteroids and other celestial material, a gradual build-up of radioactive heat probably created volcanism that eventually formed the maria. Some flows might even have been triggered by deep penetration of impacting meteorites. Dating of some volcanic lunar rocks was in harmony with this suggestion.

The lunar samples contained large amounts of glass in many forms and in a great variety of colors. It occurred as spatter or glazing on rock fragments, within breccias, as irregular blebs, and as tiny spheres, teardrops, and dumb-bells. These lunar glasses were interpreted as the result of melting and vaporization of rocks and soils caused by hypervelocity impact. Evidence of strong shock was reported in various minerals, and included formation of maskelynite, selective metamorphism, microfracturing, and deformation of lamellae and of crystal shapes.[24] E. C. T. Chao et al. remarked in 1970 that "evidence of shock metamorphism is abundant in the Apollo 11 returned lunar sample. This finding leaves no doubt that meteor impact is an important, if not the most important, crater-forming process on the surface of the moon."[25]

Other spectacular achievements of the space program involved travel by unmanned spacecraft to the vicinity of the nearer planets. Transmission to earth of the first relatively close-up images of their somewhat hostile surfaces revealed that the faces of Mercury and Mars, like that of the moon, were covered with the now-familiar form of impact craters; and that beneath its heavy cloud cover, evidence of impact craters could be discerned on the surface of Venus. The impacting of meteorites is evidently an ancient and widespread phenomenon. D. E. Gault *et al.*, in 1975, remarked that

> it is now firmly established that impact cratering has been a geologic process of primary significance in the evolution of all the terrestrial planets . . . and it is clear that a thorough understanding of cratering processes and formation is essential for gaining further insight into the early history and subsequent development of all the terrestrial planets. [26]

Similar views were expressed in 1976 by E. M. Shoemaker, in whose opinion "the impact of solid bodies is the most fundamental of all processes that have taken place on the terrestrial planets. . . . Without impact, Earth, Mars, Venus, and Mercury wouldn't exist. Collision of smaller objects is the process by which the terrestrial planets were born."[27]

And so it was that the investigation of some holes in the ground led eventually to a small increase in our meager knowledge of the astonishingly wonderful universe in which we live.

# *Notes to the Chapters*

INTRODUCTION
1. Because the metric system of measurement is used in most scientific writing, the approximate dimensions of meteorite craters throughout this book are given in metric units as well as in the more familiar U.S. Customary (or English) units, which are used for other measurements. One kilometer (km)=0.6 miles; 1 meter (m)=3.28 feet.

CHAPTER 1
1. Pliny (1938 translation by H. A. Rackham), pp. 27, 285.
2. L. Thorndike, 1958, p. 50.
3. J. Wallis, 1677, pp. 864–865.
4. Ibid.
5. E. Halley, 1714, p. 159.
6. Ibid., p. 160.
7. E. Halley, 1719, pp. 989–990.
8. J. Pringle, 1759, pp. 266–267.
9. C. Blagden, 1784, p. 222.
10. Ibid., p. 218.
11. E. Halley, 1714, p. 161–162.
12. E. C. Howard, 1802, p. 175.
13. Ibid., p. 180.
14. Ibid., p. 168.
15. Ibid., p. 187.
16. Ibid., p. 186.
17. Ibid., p. 191.
18. Ibid., p. 203.
19. A. F. Fourcroy, 1803, p. 304.
20. Ibid.
21. Both Fourcroy and Biot use the spelling "Laigle," which was evidently usual at the time they wrote. In modern writing it usually appears as "L'Aigle."
22. These comments were included in a letter from J. B. Biot to the French minister of the Interior (*Philosophical Magazine* [*Tilloch's*] vol. 16, pp. 224–226, English translation). The letter is signed "C. Biot." This, as pointed out by Cecil Schneer (personal communication), simply means "Citizen Biot."
23. D. W. Sears, 1975, pp. 215–216.
24. Very little is known about G. Thomson. See R. T. Gunther, 1939, pp. 667, 668.
25. G. Thomson, 1808, as quoted by F. A. Paneth, 1960, p. 178.
26. C. S. Smith, 1962, pp. 971, 972.
27. E. L. Krinov, 1960a, p. 196.

28. F. A. Paneth, 1960, p. 176.

29. J. N. Lockyer, 1890, p. 63.

## CHAPTER 2

1. C. E. Dutton, 1885, p. 126.

2. Ibid., p. 121.

3. G. K. Gilbert, A. R. Marvine, E. E. Howell, and J. J. Stevens, 1875, p. 538.

4. N. H. Darton, 1905; C. R. Herrick, 1900.

5. C. A. Cotton, 1944; O. D. von Engeln, 1932.

6. G. Müller and G. Veyl, 1957.

7. E. M. Shoemaker, 1962a, p. 294.

8. J. P. Bradbury, 1967.

## CHAPTER 3

1. W. G. Hoyt, 1983.

2. A. E. Foote, 1891, p. 414.

3. Surface rock at Meteor Crater is a reddish brown sandstone of the Triassic Moencopi formation. It lies as a thin and patchy veneer on sandy dolomitic limestone of the Permian Kaibab formation. The Kaibab extends downward to a depth of about 350 feet and rests on a thin bed of yellow or reddish-brown sandstone known as the Toroweap formation (Permian). Below this is the Coconino sandstone (also Permian), which is some 700 to 800 feet thick.

4. W. M. Davis, 1922, p. 183.

5. Ibid.

6. For many years the word *meteor* was used to designate a variety of atmospheric phenomena. The *Oxford English Dictionary* (1933) gives several definitions, including the following:

> A luminous body seen temporarily in the sky, and supposed to belong to a lower region than that of the heavenly bodies. In its modern restricted sense the term may be scientifically defined as meaning a small mass of matter from celestial space rendered luminous by the heat engendered by collision with the earth's atmosphere.

Thus, in early writing, and during the first half of the twentieth century up into the 1950s, the word "meteoric" was frequently used with reference to meteors in the above sense, including *both* the bright streak in the sky *and* the material which caused it. Today, in attempts to achieve greater precision, the word *meteor* in scientific writing is used to designate only the bright streak in the sky. A solid object in interplanetary space within the solar system is now called a *meteoroid,* and a *meteorite* is defined as a fragment of meteoroid material that has survived atmospheric passage and impact on a planetary surface.

It is often said that the name of the Arizona Meteor Crater is an example of the former use of the word *meteor,* and that if it were being newly named today, it would no doubt be called the Arizona Meteorite Crater. However, according to William Hoyt (1983), Meteor Crater was named in accordance with accepted rules of geological nomenclature: it is usual for a meteorite crater to be given the name of the nearest post office. Thus, the Canyon Diablo irons carried the name of the Canyon Diablo post office, located some twelve miles from the crater. A more convenient post office, only five miles from the crater, was established in 1905. It was named "Meteor, Arizona"; and although it was short-lived, the name Meteor Crater, already favored by

H. L. Fairchild, was approved. Whether Meteor, Arizona, if it were being newly named today, would instead be "Meteorite, Arizona," is another question!

7. G. K. Gilbert, 1893*b*, p. 187.
8. W. M. Davis, 1922, p. 176.
9. G. K. Gilbert, 1893*a*, p. 261 ff.
10. W. M. Davis, 1922, p. 185.
11. G. K. Gilbert, 1886, p. 286.
12. G. K. Gilbert, 1896, p. 12.
13. W. M. Davis, 1922, p. 197.
14. Ibid., p. 184.
15. R. J. Chorley, R. P. Beckinsale, and A. J. Dunn, 1973, p. 383.
16. B. C. Tilghman, 1905, p. 898.
17. G. P. Merrill, 1907, p. 550.
18. G. P. Merrill, 1908, p. 494.
19. O. C. Farrington, 1915.
20. D. M. Barringer, 1914, p. 558.
21. C. Fisher, 1943, p. 169.
22. N. B. Svensson and F. E. Wickman, 1965.
23. D. M. Barringer, 1914, p. 556-557.
24. D. M. Barringer, Jr., 1963, p. ix.
25. A. F. Rogers, 1930, p. 202.
26. F. R. Moulton, 1929*b*, second of two letters of Aug. 30, 1929, to Professor W. S. Hutchinson.
27. F. R. Moulton, 1929*c*, p. 126. These calculations were not published because, according to C. E. Gasteyer (1970), Moulton did not wish to be drawn into arguments over the size of the meteorite.
28. F. R. Moulton, 1929*c*, foreword, p. 14.
29. F. R. Moulton, 1931, p. 305.
30. J. J. Jakosky, 1932, p. 392.

## CHAPTER 4

1. B. Barringer, 1967, pp. 162, 163.
2. E. M. Sellards and G. Evans, 1941.
3. G. Evans, 1982 (personal communication).
4. F. J. W. Whipple, 1930, p. 300.
5. C. P. Skrine, 1931, p. 328.
6. L. J. Spencer, 1933*c*, p. 367.
7. W. A. Cassidy, L. M. Villar, T. E. Bunch, T. P. Kohman, and D. J. Milton, 1965.
8. W. A. Cassidy, 1970, p. 187.

## CHAPTER 5

1. M. Kerker, 1960.
2. A. C. Gifford, 1930, p. 76.
3. H. E. Ives, 1919, pp. 248, 249.
4. A. C. Gifford, 1924, p. 135.
5. Ibid., p. 139.
6. Ibid., p. 136.
7. Ibid., p. 140.
8. Ibid., p. 135.
9. A. C. Gifford, 1930, p. 75.
10. A. R. Alderman, 1932, p. 27.
11. Ibid., p. 26.
12. L. J. Spencer, 1932*b*, p. 781.

13. L. J. Spencer, 1933*a*, p. 391.
14. L. J. Spencer, 1933*c*, p. 368.
15. Ibid., p. 367.
16. L. J. Spencer, 1933*a*, p. 395.
17. L. J. Spencer, 1933*b*, p. 241.
18. L. J. Spencer, 1933*a*, p. 400.
19. L. J. Spencer, 1937, p. 655.
20. Ibid.

CHAPTER 6
1. This anecdote was kindly supplied by W. E. Elston.
2. Southern Ohio is part of a region that was once inhabited by ancient people now called Mound Builders. Confusion has sometimes arisen because a well-known archaeological site called Serpent Mound is not far from the geologically complicated hill, also called Serpent Mound, which was investigated by W. H. Bucher.
3. W. H. Bucher, 1936, p. 1063. Bucher's list of cryptovolcanic structures in the United States was first given at the International Geological Conference in 1933, but it was published in 1936. It was immediately referred to, even before 1936, as Bucher's 1933 report. There is now some confusion in the literature, as many writers still refer to it as Bucher's 1933 report; but some, especially later writers, refer to it with a date of 1936, as I have done here.
4. Ibid.
5. W. H. Bucher, 1925, p. 195.
6. W. H. Bucher, 1936, p. 1059.
7. W. H. Bucher, 1925, p. 226.
8. Ibid., p. 227.
9. Ibid., p. 231.
10. Ibid., p. 232.
11. A. A. Baker, 1933.
12. H. W. C. Prommel and H. E. Crum, 1927.
13. T. S. Harrison, 1927, p. 128.
14. This interpretation of Upheaval Dome was widely accepted for many years. It was challenged in 1983, however, by E. M. Shoemaker and K. E. Herkenhoff. Their investigations revealed peculiar faults and other structural features which convinced them that this puzzling uplift is the deeply eroded scar of meteoritic impact.
15. W. H. Bucher, 1936, p. 1068.
16. Ibid., p. 1070.
17. Ibid., p. 1071.
18. J. Collett, 1883, pp. 58–59.
19. S. S. Gorby, 1886, p. 240.
20. Ibid., p. 237.
21. M. Thompson, 1892, p. 185.
22. Ibid., p. 180.
23. M. N. Elrod and A. C. Benedict, 1891.
24. G. H. Ashley, 1898, p. 190.
25. L. C. Ward, 1906, p. 214.
26. E. M. Kindle, 1903.
27. E. R. Cumings, 1922, p. 449.
28. E. R. Cumings and R. R. Shrock, 1928, p. 580.
29. R. R. Shrock and C. A. Malott, 1933, p. 339.

30.  W. H. Bucher, 1936, p. 1074.

31.  K. E. Born and C. W. Wilson, 1939.

32.  P. B. King, 1930, pp. 123–125.

33.  H. P. T. Rohleder, 1933, p. 494.

34.  J. D. Boon and C. C. Albritton, Jr., 1936, p. 7.

35.  J. D. Boon and C. C. Albritton, 1937, p. 56.

36.  Ibid., p. 53.

37.  J. D. Boon and C. C. Albritton, Jr., 1938, p.47.

CHAPTER 7

1.  A. M. Miller, 1923, p. 435.

2.  Much of the information concerning Nininger's early life is taken from his book *Find a Falling Star*, published in 1972.

3.  H. H. Nininger, 1972, p. 13.

4.  Ibid., p. 20.

5.  Ibid., p. 33.

6.  F. L. Whipple, 1972.

7.  There has been some confusion of Mrs. Kimberly's name with that of her daughter Mary. Dr. Nininger's son-in-law, Glenn Huss, informed me in 1984 that he had settled the matter by visiting Mrs. Kimberly's grave, where her tombstone bears the name Eliza.

8.  G. F. Kunz, 1890, p. 316.

9.  H. H. Nininger and J. D. Figgins, 1933, p.10.

10.  H. H. Nininger, 1933, p. 65.

11.  H. H. Nininger, 1972, p. 108.

12.  Ibid., p. 101.

13.  Ibid., p. 175.

14.  Ibid., p. 176.

15.  Ibid., p. 179.

16.  Ibid., p. 181.

17.  J. S. Rinehart, 1957, p. 96.

18.  B. C. Tilghman, 1905, p. 898.

CHAPTER 8

1.  H. H. Nininger and G. I. Huss, 1961.

2.  C. T. Madigan, 1937, 1940.

3.  F. Reeves and R. O. Chalmers, 1949.

4.  D. J. Guppy and R. S. Matheson, 1950.

5.  W. Cassidy, 1954, p. 197.

6.  T. Monod and A. Pourquié, 1951.

7.  E. C. T. Chao, E. J. Dwornik, and C. W. Merrill, 1966.

8.  R. S. Dietz, R. Fudali, and W. Cassidy, 1969.

9.  B. M. French, J. B. Hartung, N. M. Short, and R. S. Dietz, 1970.

10.  T. Monod, 1955, p. 309.

11.  Ibid., p. 310.

12.  L. and J. LaPaz, 1954, pp. 192, 195.

13.  P. M. Millman, 1956, p. 63.

14.  V. B. Meen, 1950, p. 175.

15.  Ibid., p. 177.

16.  J. A. C. Keefe, S. H. Ward, and J. T. Wilson, 1957, pp. 151–152.

17.  P. M. Millman, 1956, p. 64.

18.  C. L. Janssen, 1951.

19.  D. Hoffleit, 1952, p. 8.

20. E. L. Krinov, 1966, p. 279.
21. Ibid., p. 285.
22. L. J. Spencer, 1933*b*, p. 228.


CHAPTER 9
    1. J. Green, 1965, p. 390.
    2. R. A. Proctor, 1873, p. 345.
    3. T. J. J. See, 1910, p. 20.
    4. D. P. Beard, 1917, p. 176.
    5. D. M. Barringer, 1914.
    6. E. Thomson, 1914, p. 38.
    7. A. Wegener, 1921 (1975 translation), pp. 213, 215, 221, 234.
    8. L. J. Spencer, 1937, p. 656.
    9. R. S. Dietz, 1946*b*, p. 465.
   10. R. B. Baldwin, 1978, p. 371.
   11. R. B. Baldwin, 1949, p. 114.
   12. Ibid., p. 153.
   13. Ibid., p. 113.


CHAPTER 10
    1. R. A. Daly, 1947, p. 125.
    2. S. J. Shand, 1916, p. 217.
    3. R. A. Daly, 1947, p. 138.
    4. R. S. Dietz, 1946*b*, p. 466.
    5. R. S. Dietz, 1947.
    6. R. S. Dietz, 1959.
    7. N. H. Darton, 1945, p. 1154.
    8. D. Hager, 1953, p. 821.
    9. D. M. Barringer, 1926.
   10. D. Hager, 1953, p. 821.
   11. E. M. Shoemaker, 1981 (personal communication).
   12. G. J. F. MacDonald, 1956.
   13. H. H. Nininger, 1956, p. 50.
   14. Ibid., p. 140.
   15. W. T. Pecora, 1960.
   16. By definition, a mineral is a naturally occurring material. Coesite was therefore not classed as a mineral until it was first discovered occurring naturally, at Meteor Crater.
   17. E. C. T. Chao, E. M. Shoemaker, and B. Madsen, 1960.
   18. E. M. Shoemaker and E. C. T. Chao, 1961.
   19. E. C. T. Chao, J. J. Fahey, and J. Littler, 1961.
   20. H. P. T. Rohleder, 1936, p. 53.
   21. J. Littler, J. J. Fahey, R. S. Dietz, and E. C. T. Chao, 1961.
   22. A. J. Cohen, T. E. Bunch, and A. M. Reid, 1961.
   23. T. E. Bunch and A. J. Cohen, 1961, p. 380.
   24. E. M. Shoemaker, 1981 (personal communication).
   25. E. C. T. Chao, J. J. Fahey, J. Littler, and D. J. Milton, 1962.
   26. D. M. Barringer, 1909, p. 6.
   27. R. S. Dietz, 1947.
   28. E. M. Shoemaker, D. E. Gault, and R. V. Lugn, 1961, p. D-365.
   29. Ibid.
   30. H. J. Milledge, 1961; Y. L. Orlov, 1973.

CHAPTER 11
1. V. B. Meen, 1957.
2. Ibid., p. 65.
3. P. M. Millman, B. A. Liberty, J. F. Clark, P. L. Willmore, and M. J. S. Innes, 1960, p. 4.
4. Ibid., p. 6.
5. Ibid.
6. Seismic methods of subsurface exploration involve determination of the speed of sound through rocks. For this purpose, sound waves are created by a small explosion. Sensitive instruments then determine the speed at which the sound waves travel through the rock, as well as any refraction or reflection which they may undergo. "Travel times" of sound through various types of rock have been experimentally determined. Thus, the speed of the sound indicates the type of rock through which it is passing, and a change of speed is interpreted as a change of rock type. The location of an abrupt change in the rock can often be determined by reflection of some of the sound waves.
7. Diamond drilling involves the use of a rotary drilling bit studded with small industrial-type diamonds which, because of their hardness, can cut through most subsurface materials. The cylindrical core thus created provides complete information concerning the underlying rocks.
8. M. J. S. Innes, 1964.
9. C. S. Beals, M. J. S. Innes, and J. A. Rottenberg, 1960, p. 212; C. S. Beals, M. J. S. Innes, and J. A. Rottenberg, 1963, p. 246.
10. T. E. Bunch and A. J. Cohen, 1961.
11. P. G. Downes, 1943, p. 65–66.
12. M. J. S. Innes, W. J. Pearson, and J. W. Geuer, 1964, p. 23.
13. C. S. Beals, G. M. Ferguson, and A. Landeau, 1956b, No. 6, p. 260.
14. P. M. Healy, L. LaPaz, and F. C. Leonard, 1953, p. 160.
15. C. S. Beals, M. J. S. Innes, and J. A. Rottenberg, 1960, p. 207; C. S. Beals, M. J. S. Innes, and J. A. Rottenberg, 1963, p. 238.
16. C. S. Beals, M. J. S. Innes, and J. A. Rottenberg, 1963, p. 277.
17. K. E. Bullen, 1953, p. 84.
18. C. S. Beals, M.J. S. Innes, and J. A. Rottenberg, 1960, p. 261.
19. In U.S.G.S. Professional Paper #400, published in 1960, E. M. Shoemaker refers to his article "Impact Mechanics at Meteor Crater" as being already in press. It was published in 1963 in Middlehurst and Kuiper's *The Moon, Meteorites, and Comets.* Presumably the 1963 version of C. S. Beals *et al.* was also in press in 1960.
20. T. E. Bunch and A. J. Cohen, 1961.
21. E. M. Shoemaker, 1962a, p. 284.
22. C. T. Hardy, 1953, p. 2580.
23. E. M. Shoemaker, 1963, p. 302.
24. Ibid., p. 312.
25. J. J. Gilvarry and J. E. Hill, 1956.

CHAPTER 12
1. W. H. Bucher, 1963b, p. 599.
2. Ibid., p. 598.
3. Ibid., p. 600.
4. Ibid., p. 614.
5. W. H. Bucher, 1963b, p. 609.
6. W. H. Bucher, 1963a, p. 1242.
7. Ibid., p. 1243.

8. R. S. Dietz, 1963*a*, p. 650.

9. Ibid., p. 654.

10. Ibid.

11. Ibid., p. 655.

12. Ibid., p. 663.

13. This and subsequent extracts from Bucher's correspondence were kindly supplied by W. E. Elston. Used with permission from the Department of Geological Sciences, Columbia University.

14. Extract from a letter to W. H. Bucher from C. S. Beals (1964). Kindly supplied by W. E. Elston, and used with permission of Mrs. C. S. Beals.

15. Extract from a letter from W. H. Bucher to W. E. Elston (1964).

16. W. E. Elston, 1979, personal communication.

17. Extract from a letter from W. H. Bucher to W. E. Elston (1964).

18. W. E. Elston, 1978, personal communication.

19. C. S. Beals, M. J. S. Innes, and J. A. Rottenberg, 1960, p. 261.

20. G. J. H. McCall, 1964, p. 253.

21. W. H. Bucher, 1963*b*, p. 619.

22. F. G. Snyder and P. E. Gerdemann, 1965, p. 465.

23. G. C. Amstutz, 1964, p. 352.

24. K. L. Currie, 1964*b*, p. 93.

25. K. L. Currie, 1965, p. 918.

26. Ibid., p. 928.

27. C. S. Beals, I. Halliday, and J. T. Wilson, 1968, p. 1.

28. Ibid., p. 2.

29. M. J. S. Innes, 1957.

30. D. B. McIntyre, 1962, p. 1647.

31. M. R. Dence, 1965, p. 955.

32. M. R. Dence, 1968, p. 181.

33. P. B. Robertson, 1968.

34. Ibid., p. 106.

CHAPTER 13

1. M. R. Dence, 1964.

2. G. P. Merrill, 1907, p. 538.

3. J. Perkins, 1820.

4. M. C. Demler, 1961.

5. F. Birch, 1963, p. 138.

6. M. J. Buerger, 1930, p. 45.

7. E. B. Knopf, 1933, p. 470.

8. D. Griggs and J. F. Bell, 1938, p. 1724.

9. P. W. Bridgman, 1936, p. 669.

10. An important series of experiments on the Yule marble included work by the following: D. Griggs and W. B. Miller, 1951; J. W. Handin and D. Griggs, 1951; F. J. Turner and C. S. Ch'ih, 1951; D. Griggs, F. J. Turner, I. Borg, and J. Sosoka, 1951; D. Griggs, F. J. Turner, I. Borg, and J. Sosoka, 1953; I. Borg and F. J. Turner, 1953; F. J. Turner, D. Griggs, R. H. Clark, and R. H. Dixon, 1956.

11. J. M. Christie, D. T. Griggs, and N. L. Carter, 1964, p. 755.

12. R. Berman and F. Simon, 1955, p. 333.

13. Ibid., p. 338.

14. H. H. Uhlig, 1954.

15. M. E. Lipschutz and E. Anders, 1961, p. 91.

16. E. Anders, 1965, p. 32.

17. H. Urey, 1956*a*, J. F. Lovering, 1957; A. E. Ringwood, 1960.

18. E. Anders, 1961.
19 M. E. Lipschutz and E. Anders, 1961, p. 100.
2C. M. E. Lipschutz, 1962, p. 1267.
21. A. P. Vinogradov and G. P. Vdovykin, 1963.
22. M. E. Lipschutz, 1964, p. 1434.
23. N. L. Carter and G. C. Kennedy, 1964; N. L. Carter and G. C. Kennedy, 1966; E. Anders and M. E. Lipschutz, 1966.
24. W. v. Engelhardt and F. Hörz, 1965; F. Hörz, 1965.
25. A. El Goresy, 1965.
26. D. Cummings, 1965.
27. M. R. Dence, 1968.

CHAPTER 14
1. A. L. Hall, 1932, p. 152.
2. R. A. Daly and G. A. F. Molengraaff, 1924, p. 2.
3. A. L. Hall, 1932, p. 141.
4. R. A. Daly and G. A. F. Molengraaff, 1924, p. 2.
5. A. L. Hall, 1932, p. 17.
6. R. A. Daly, 1928, p. 723.
7. Ibid., p. 755.
8. Ibid., p. 756.
9. A. L. Hall, 1932, p. 388.
10. Ibid., p. 393.
11. H. Kynaston, 1910b, p. 25.
12. Pilandsberg is the officially recognized spelling. Shand pointed out, however, that the name originated as Pilaan's Berg, and he preferred to omit the d.
13. S. J. Shand, 1930, p. 415.
14. A. L. Hall, 1932, p. 431.
15. A. L. Hall and G. A. F. Molengraaff, 1925, p. 145.
16. Ibid., p. 160.
17. Ibid., p. 142.
18. Ibid., p. 132.
19. A. L. Hall, 1932, p. 450.
20. S. J. Shand, 1916, p. 200.
21. Ibid., p. 208.
22. Ibid., p. 215.
23. Ibid., pp. 216–217.
24. Ibid., p. 219.
25. A. L. Hall and G. A. F. Molengraaff, 1925, p. 108.
26. Ibid., p. 156.
27. A. L. Hall, 1932, p. 89.
28. Ibid., p. 17.
29. Ibid., p. 18.
30. P. B. King, 1930, p. 125.
31. J. D. Boon and C. C. Albritton, Jr., 1937, p. 63.
32. J. D. Boon and C. C. Albritton, Jr., 1936.
33. J. D. Boon and C. C. Albritton, Jr., 1937, p. 64.
34. J. D. Boon and C. C. Albritton, Jr., 1938, p. 53.
35. D. W. Bishopp, 1941, p. 9.
36. R. A. Daly, 1947, p. 137.
37. Ibid.
38. W. E. Elston (personal communication).
39. A. du Toit, 1954, p. 200.

40. E. M. Shoemaker, 1960*a*, p. 170.
41. R. B. Hargraves, 1961, p. 152.
42. R. S. Dietz, 1961, p. 499.
43. A. Poldervaart, 1962, p. 244.
44. R. S. Dietz, 1961, p. 515.
45. W. I. Manton, 1965, p. 1047.
46. Ibid., p. 1049.
47. J. R. Smyth and C. J. Hatton, 1977, p. 289.
48. W. Hamilton, 1970, p. 372.
49. Ibid., p. 371.
50. R. C. Rhodes, 1975, p. 550.
51. M. R. Dence, 1964, p. 259.

CHAPTER 15
1. A. P. Coleman, 1905, p. 168.
2. Ibid., p. 169.
3. Ibid., p. 170.
4. A. P. Coleman, 1907, p. 759.
5. Ibid., p. 760.
6. A. P. Coleman, 1905, p. 76.
7. G. H. Williams, 1891, p. 138.
8. T. L. Walker, 1897, p. 42.
9. Ibid., p. 44.
10. W. H. Collins, 1937, p. 15.
11. Because of controversy about the origin, emplacement, and sub-surface shape of the igneous sheet, the term *irruptive,* which carries no genetic implications, was introduced by Collins, and gradually replaced the term *eruptive.*
12. W. H. Collins, 1934, p. 176.
13. Ibid., p. 161.
14. W. H. Colllins, 1937, p. 29.
15. W. H. Collins and E. D. Kindle, 1935, p. 44.
16. C. W. Knight, 1916, p. 812.
17. W. H. Collins, 1937, p. 16.
18. W. H. Collins, 1936, p. 29.
19. Ibid., p. 43.
20. Ibid., p. 41.
21. W. H. Collins, 1937, p. 28.
22. H. W. Fairbairn and G. M. Robson, 1942, p. 6.
23. E. C. Speers, 1957, p. 513.
24. J. E. Thomson, 1957, p. 109.
25. Ibid., p. 112.
26. Ibid., p. 110.
27. W. Hamilton, 1961, p. 437.
28. Ibid., p. 445.
29. J. E. Hawley, 1962, p. v.
30. Ibid., p. 3.
31. Ibid., p. xiii.
32. R. S. Dietz, 1964, p. 420.
33. Ibid., p. 424.
34. Ibid., p. 432.
35. R. S. Dietz and L. W. Butler, 1964, p. 281.
36. J. Guy Bray and Geological Staff, 1966, p. 245.
37. T. L. Walker, 1897, p. 45.

38. A. P. Coleman, 1905, p. 131.
39. B. M. French, 1967, 1968.
40. B. J. Skinner and J. J. Fahey, 1963.
41. K. D. Card, 1964.
42. B. M. French, 1968, p. 406.
43. J. V. Guy Bray, 1972, p. 2.
44. R. S. Dietz, 1972, pp. 29, 35.
45. M. R. Dence and J. Popelar, 1972, p. 117.
46. M. R. Dence, P. B. Robertson, and R. L. Wirthlin, 1974.
47. M. R. Dence and J. Popelar, 1972, p. 123.

## CHAPTER 16
1. B. M. French, 1970, p. 468.
2. M. R. Dence, 1964, 1965.
3. V. L. Masaitis, S. I. Futergendler, and M. A. Gnevushev, 1972; V. L. Masaitis, M. V. Mikhailov, and T. V. Selivanovskaya, 1971.
4. G. Kürat and W. Richter, 1969.
5. D. J. Milton and C. W. Naesser, 1971.
6. R. F. Fudali, D. P. Gold, and J. J. Gurney, 1973.
7. H. E. LeGrand, 1953.
8. C. Darwin, 1844 (1896 ed.), p. 44.
9. L. J. Spencer, 1933e, p. 571.
10. Ibid.
11. L. J. Spencer, 1933d, p. 117.
12. H. C. Urey, 1962, p. 747.
13. J. S. Rinehart, 1958.
14. V. E. Barnes, 1964.
15. W. Gentner, H. J. Lippolt, and O. A. Schaeffer, 1963.
16. W. Gentner, H. J. Lippolt, and O. Müller, 1964.
17. R. L. Fleischer and P. B. Price, 1964; R. L. Fleischer, P. B. Price, and R. M. Walker, 1965; C. C. Schnetzler, W. H. Pinson, and P. M. Hurley, 1966.
18. A. L. Turkevich, E. J. Franzgrote, and J. H. Patterson, 1969; J. H. Patterson et al., 1970.
19. R. F. Fudali and R. J. Ford, 1979, p. 292.
20. R. S. Dietz, 1977.
21. D. Storzer and G. A. Wagner, 1979.
22. Z. Kopal, 1971, p. 5.
23. E. M. Shoemaker, 1962b, p. 109.
24. M. R. Dence et al., 1970.
25. E. C. T. Chao et al., 1970, p. 311.
26. D. E. Gault et al., 1975, p. 2444.
27. E. M. Shoemaker, 1976, p. 1.

# References

Abercrombie, T. J. 1966. "Saudi Arabia: Beyond the Sands of Mecca." *Nat. Geog. Mag.* (Jan.), p. 35.

Adams, F. D., and J. T. Nicolson. 1901. "An Experimental Investigation into the Flow of Marble." *Trans. Roy. Soc. London,* Ser. A., vol. 195, pp. 363–401.

Alderman, A. R. 1932. "Meteorite Craters at Henbury, Central Australia." *Mineralog. Mag.* vol. 23, pp. 19–32.

Allan, D. W., and J. A. Jacobs. 1956. "The Melting of Asteroids and the Origin of Meteorites." *Geochim. et Cosmochim. Acta,* vol. 9, pp. 256–272.

Amstutz, G. C. 1964. "Impact, Cryptoexplosion, or Diapiric Movements?" *Trans. Kansas Acad. Sci.,* vol. 67, pp. 343–356.

Anders, E. 1961. "Comments on the Origin of Natural Diamonds." *Astrophys. Jour.,* vol. 134, p. 1006.

———. 1965. "Diamonds in Meteorites." *Sci. Am.,* vol. 213, pp. 26–36.

Anders, E., and M. E. Lipschutz. 1966. "Critique of Paper by N. L. Carter and G. C. Kennedy: 'Origin of Diamonds in the Canyon Diablo and Novo Urei Meteorites.'" *Jour. Geophys. Res.,* vol. 71, pp. 643–661. Also "Reply," pp. 673–674.

Ashley, G. H. 1898. "Coal Deposits of Indiana." Indiana Dept. Coal and Nat. Res., *23rd Ann. Rept.,* p. 190.

Bailey, E. B. 1926. "Domes in Scotland and South Africa: Arran and Vredefort." *Geol. Mag.,* vol. 63, pp. 481–495.

Baker, A. A. 1933. "Geology and Oil Possibilities of the Moab District, Grand and San Juan Counties, Utah." *U.S.G.S. Bull. No. 841,* p. 81.

Baldwin, R. B. 1942. "The Meteoritic Origin of Lunar Craters." *Pop. Astron.,* vol. 50, pp. 365–369.

———. 1943. "The Meteoritic Origin of Lunar Structures." *Pop. Astron.,* vol. 51, pp. 117–127.

———. 1949. *The Face of the Moon.* Univ. Chicago Press.

———. 1978. "An Overview of Impact Cratering." *Meteoritics,* vol. 13, pp. 364–379.

Barlow, A. E. 1906. "On the Origin and Relations of the Nickel and Copper Deposits of Sudbury, Ontario, Canada." *Econ. Geol.,* vol. 1, 1905, pp. 454–475 and 545–553.

Barnes, V. E. 1940. "Contributions to Geology, 1939, Part 2, North American Tektites." *Univ. of Texas Pubn. 3945,* pp. 477–582 (reprinted in *Tektites*, V. E. and M. A. Barnes, 1973, Dowden, Hutchinson & Ross, Inc., Stroudsberg, Pa.).

————. 1963. "Detrital Mineral Grains in Tektites." *Science*, vol. 142, pp. 1651–1652.

————. 1964. "Terrestrial Implication of Layering, Bubble Shape, and Minerals along Faults in Tektite Origin." *Geochim. et Cosmochim. Acta*, vol. 28, pp. 1267–1271.

Barringer, B. 1964. "Daniel Moreau Barringer (1860–1929) and His Crater." *Meteoritics*, vol. 2, pp. 183–199.

————. 1967. "Historical Notes on the Odessa Meteorite Crater." *Meteoritics*, vol. 3, pp. 161–168.

Barringer, D. M. 1905. "Coon Mountain and Its Crater." *Proc. Acad. Nat. Sci. Phil.*, vol. 57, pp. 861–886.

————. 1909. "Meteor Crater (Formerly Called Coon Mountain or Coon Butte) in N. Central Arizona." Paper read before Nat. Acad. Sci., Princeton Univ. Privately printed.

————. 1914. "Further Notes on Meteor Crater." *Proc. Acad. Nat. Sci. Phil.*, vol. 76, pp. 556–565.

————. 1926. "Craters and Sink-holes" (reply to letter by Dorsey Hager). *Eng. and Min. Journal-Press*, vol. 121, p. 450.

Barringer, D. M., Jr. 1928. "A New Meteorite Crater." *Proc. Acad. Nat. Sci. Phil.*, vol. 80, pp. 307–311.

————. 1963. Quoted in "Daniel Moreau Barringer 1900–1962." *Meteoritics*, vol. 2, p. ix.

Beals, C. S. 1957. "A Probable Meteorite Crater of Great Age." *Sky and Telescope*, vol. 16, pp. 526–528.

————. 1958. "Fossil Meteorite Craters." *Sci. Am.* (July), pp. 32–39.

————. 1960. "A Probable Meteorite Crater of Precambrian Age at Holleford, Ontario." *Pubn. of the Dom. Observ., Ottawa*, vol. 24., no. 6.

————. 1964. "A Re-examination of the Craters in the Faugeres-Cabrerolles Region of Southern France." *Meteoritics*, vol. 2, no. 2, pp. 85–91.

Beals, C. S., G. M. Ferguson, and A. Landau. 1956*a*. "The Holleford Crater in Ontario." *Sky and Telescope*, vol. 15, no. 7, p. 296.

————. 1956*b*. "A Search for Analogies Between Lunar and Terrestrial Topography on Photographs of the Canadian Shield, Parts I and II." *Roy. Astron. Soc. Can. Jour.*, vol. 50, nos. 5 and 6, pp. 203–211 and 250–261.

Beals, C. S., and I. Halliday. 1965. "Impact Craters of the Earth and Moon." *Roy. Astron. Soc. Can. Jour.*, vol. 59, no. 5, pp. 199–216.

Beals, C. S., I. Halliday, and J. T. Wilson. 1968. "Theories of the Origin of Hudson Bay." In *Science, History and Hudson Bay*, Dept. of Energy, Mines, and Resources, Ottawa, Canada, pp. 1–49.

Beals, C. S., M. J. S. Innes, and J. A. Rottenberg. 1960. "The Search for Fossil Meteorite Craters, I and II." *Current Science*, vol. 29, pp. 205–218 and 249–262.

————. 1963. "Fossil Meteorite Craters." In *The Moon, Meteorites, and Comets,* edited by B. M. Middlehurst and G. P. Kuiper, pp. 235–284. Univ. Chicago Press.

Beard, D. P. 1917. "The Impact Origin of the Moon's Craters." *Pop. Astron.,* vol. 25, pp. 167–177.

Beer, W., and J. H. Mädler. 1837. *Der Mond . . . .* Simon Schropp, Berlin.

Bell, R. 1891. "The Nickel and Copper Deposits of Sudbury District, Canada." *Bull. Geol. Soc. Am.,* vol. 2, pp. 125–140.

Berman, R., and F. Simon. 1955. "On the Graphite-Diamond Equilibrium." *Zeitschrift für Elektrochemie,* vol. 59, no. 5, pp. 333–338.

Bibbins, A. B. 1926. Letter. *Eng. and Min. Journal-Press,* vol. 121, p. 932.

Biot, J. B. 1803. "Account of a Fire-ball Which Fell in the Neighbourhood of Laigle: In a Letter to the French Minister of the Interior." *Phil. Mag. (Tilloch's),* vol. 16, pp. 224–228.

Birch, F. 1963. "Some Geophysical Applications of High Pressure Research." In *Solids Under Pressure,* edited by William Paul and Douglas M. Warschauer, pp. 137–162. McGraw–Hill Book Co., New York.

Bishopp, D. W. 1941a. "The Geodynamics of the Vredefort Dome." *Trans. Geol. Soc. South Africa,* vol. 44, pp. 1–18.

———. 1941b. "Remarks in Discussion of a Paper by D. W. Bishopp." *Trans. Geol. Soc. South Africa,* pp. xci–xcvii.

Bjork, R. L. 1961. "Analysis of the Formation of Meteor Crater, Arizona: A Preliminary Report." *Jour. Geophys. Res.,* vol. 66, pp. 3379–3387.

Blackwelder, E. 1953. "Discussion: Crater Mound—Meteor Crater." *Bull. Am. Assoc. Petrol. Geol.,* vol. 37, pp. 2577–2580.

Blagden, C. 1784. "An Account of some Late Fiery Meteors; with Observations etc." *Phil. Trans. Roy. Soc.,* vol. 74, pp. 201–232.

Bonney, T. G. 1888. "Notes on a Part of the Huronian Series in the Neighbourhood of Sudbury (Canada)." *Geol. Soc. London Quart. Jour.,* vol. 44, no. 1, pp. 32–45.

Boon, J. D., and C. C. Albritton, Jr. 1936. "Meteorite Craters and Their Possible Relationship to 'Cryptovolcanic Structures.'" *Field and Lab.,* vol. 5, no. 1, pp. 1–9.

———. 1937. "Meteorite Scars in Ancient Rocks." *Field and Lab.,* vol. 5, pp. 53–64.

———. 1938. "Established and Supposed Examples of Meteoritic Craters and Structures." *Field and Lab.,* vol. 6, pp. 44–56.

Borg, I., and F. J. Turner. 1953. "Deformation of Yule Marble: Part VI. Identity and Significance of Deformation Lamellae and Partings in Calcite Grains." *Bull. Geol. Soc. Am.,* vol. 64, pp. 1343–1352.

Born, K. E., and C. W. Wilson, Jr. 1939. "The Howell Structure, Lincoln County, Tennessee." *Jour. of Geol.,* vol. 47, pp. 371–388.

Bovenkerk, H. P., F. P. Bundy, H. T. Hall, H. M. Strong, and R. H. Wentorf, Jr. 1959. "Preparation of Diamond." *Nature,* vol. 184, pp. 1094–1098.

Boyd, F. R., and J. L. England. 1960. "The Quartz-Coesite Transition." *Jour. Geophys. Res.*, vol. 65, pp. 749–756.

Bradbury, J. P. 1967. Paleolimnology and Limnology of Zuni Salt Lake Maar. Ph.D. dissertation, Univ. New Mexico.

Branco, (Branca), W., and E. Fraas. 1905. "Das Kryptovulcanische Becken von Steinheim." *K. Preuss, Akad. Wiss. Berlin.* Abh. 1, pp. 1–64.

Branner, J. C. 1906. "The Geology of Coon Butte, Arizona: Messrs. D. M. Barringer and B. C. Tilghman." *Science* (new ser.), vol. 24, pp. 370–371.

Bray, J. V. Guy. 1972. "Introduction." *Geol. Assoc. of Canada Special Paper No. 10.*

Bray, J. V. Guy, and Geological Staff. 1966. "Shatter Cones at Sudbury." *Jour. Geol.*, vol. 74, pp. 243–245.

Breternitz, D. A. 1967. "Eruption(s) of Sunset Crater: Dating and Effects." *Plateau*, vol. 40, pp. 72–76.

Bridgman, P. W. 1931. *The Physics of High Pressure.* Reprint. Dover, 1970.

———. 1936. "Shearing Phenomena at High Pressure of Possible Importance for Geology." *Jour. Geol.*, vol. 44, pp. 653–669.

Brock, B. B. 1965. "Discussion of Paper by W. I. Manton." *Annals N.Y. Acad. Sci.*, vol. 123, pp. 1048–1049.

Brookins, D. G. 1969. "Investigation of Winkler Crater, Kansas." *Meteoritics*, vol. 4., pp. 263–264.

Brown, P. L. 1973. *Comets, Meteorites, and Men.* Robert Hale & Co., London.

Brush, S. G. 1976. "John James Waterston." In *Dict. Sci. Biography,* C. C. Gillispie, ed. Charles Scribner's Sons, New York.

Bucher, W. H. 1921. "Cryptovolcanic Structure in Ohio of the Type of the Steinheim Basin" (abs.). *Bull. Geol. Soc. Am.*, vol. 32, p. 74.

———. 1925. "Geology of Jeptha Knob, Kentucky." *Geol. Surv.*, ser. 6, vol. 21, pp. 193–237.

———. 1928. "Cryptovolcanic Regions." *Jour. Wash. Acad. Sci.*, vol. 18, pp. 521–524.

———. 1936. "Cryptovolcanic Structures in the United States." *Internat. Geol. Congr.*, 16th, Washington, D. C. 1933, vol. 2, pp. 1055–1083.

———. 1963a. "Are Cryptovolcanic Structures Due to Meteoritic Impact?" *Nature*, March 30, pp. 1241–1245.

———. 1963b. "Cryptoexplosion Structures Caused from Without or Within the Earth? ('Astroblemes' or 'Geoblemes'?)" *Am. Jour. Sci.*, vol. 261, pp. 597–649.

———. 1965. "The Largest So-called Meteorite Scars in Three Continents as Demonstrably Tied to Major Terrestrial Structures." *Annals N.Y. Acad. Sci.*, vol. 123, pp. 897–903.

Buerger, M. J. 1928. "The Plastic Deformation of Ore Minerals." *Am. Mineralogist*, vol. 13, pp. 1–17 and 35–51.

———. 1930. "Translation-gliding in Crystals." *Am. Mineralogist*, vol. 15, pp. 45–64.

Bullen, K. E. 1953. *An Introduction to the Theory of Seismology* (2nd ed.). The University Press, Cambridge, England.

Bunch, T. E., and A. J. Cohen, 1961. "Precambrian Coesite" (abs.). *Jour. Geophys Res.*, vol. 67, p. 1630a. (Presented at first western meeting, Am. Geophys. Union, Los Angeles, Dec., 1961.) Published as "Coesite and Shocked Quartz from Holleford Crater, Ontario, Canada." *Science*, Oct. 18, 1963, pp. 379–381.

———. 1964. "Shock Deformation of Quartz from Two Meteorite Craters." *Bull. Geol. Soc. Am.*, vol. 75, pp. 1263–1266.

Bunch, T. E., A. J. Cohen, and M. R. Dence. 1967. "Natural Terrestrial Maskelynite." *Am. Mineralogist*, vol. 52, pp. 244–253.

Bundy, F. P., H. T. Hall, H. M. Strong, and R. H. Wentorf. 1955. "Man-made Diamonds." *Nature*, vol. 176, pp. 51–55.

Burrows, A. G., and H. C. Rickaby, 1930. "Sudbury Basin Area." *Ann. Rept. 38*, part 3. *Ontario Dept. Mines*.

Cailleux, A., A. Guillemaut, and C. Pomeroi, 1964. "Présence de coesite indice de hautes pressions, dans l'accident circulaire des Richât (Adrar mauritanien)." *C. R. Acad. Sci. Paris*, tome 258, groupe 9, 5488–5489.

Card, K. D. 1964. "Metamorphism in the Agnew Lake Area, Sudbury District, Ontario, Canada." *Bull. Geol. Soc. Am.*, vol. 75, pp. 1011–1030.

Carr, W. K. 1952. "Ungava Crater from the Air." *Sky and Telescope* (January), pp. 61–62.

Carter, N. L. 1965. "Basal Quartz Deformation Lamellae—A Criterion for Recognition of Impactites." *Am. Jour. Sci.*, vol. 263, pp. 786–806.

Carter, N.L., J. M. Christie, and D. T. Griggs. 1961. "Experimentally Produced Deformation Lamellas and Other Structures in Quartz Sand" (abs.). *Jour. Geophys. Res.*, vol. 66, pp. 2518–2519.

———. 1964. "Experimental Deformation and Recrystallization of Quartz." *Jour. Geol.*, vol. 72, pp. 687–733.

Carter, N. L., and M. Friedman. 1965. "Dynamic Analysis of Deformed Quartz and Calcite from the Dry Creek Ridge Anticline, Montana. *Am. Jour. Sci.*, vol. 263, pp. 747–785.

Carter, N. L., and G. C. Kennedy. 1964. "Origin of Diamonds in the Canyon Diablo and Novo Urei Meteorites." *Jour. Geophys. Res.*, vol. 69, pp. 2403–2421.

———. 1966. "Origin of Diamonds in the Canyon Diablo and Novo Urei Meteorites—A Reply." *Jour. Geophys. Res.*, vol. 71, pp. 663–672.

Cassidy, W. A. 1954. "The Wolf Creek, Western Australia, Meteorite Crater (CN=−1278,192)." *Meteoritics*, vol. 1, pp. 197–199.

———. 1970. "Discovery of a New Multiton Meteorite at Campo del Cielo." *Meteoritics*, vol. 5, p. 187.

Cassidy, W. A., L. M. Villar, T. E. Bunch, T. P. Kohman, and D. J. Milton. 1965. "Meteorites and Craters of Campo Del Cielo, Argentina." *Science*, vol. 149, pp. 1055–1064.

Chao, E. C. T. 1967a. "Impact Metamorphism." In *Researches in Geochemistry*, vol. 2, pp. 204–233, edited by P. H. Abelson. John Wiley & Sons, New York.

———. 1967b. "Shock Effects in Certain Rock-forming Minerals." *Science*, vol. 156, pp. 192–202.

Chao, E. C. T., I. Adler, E. J. Dwornick, and J. Littler. 1962. "Metallic Spherules in Tektites from Isabela, Philippine Islands." *Science*, vol. 135, pp. 97–98.

Chao, E. C. T., E. J. Dwornik, and C. W. Merrill. 1966. "Nickel-Iron Spherules from Aouelloul Glass." *Science*, vol. 154, pp. 759–761.

Chao, E. C. T., J. J. Fahey, and J. Littler. 1961. "Coesite from Wabar Crater, Near Al Hadida, Arabia." *Science*, vol. 133, pp. 882–883.

Chao, E. C. T., J. J. Fahey, J. Littler, and D. J. Milton. 1962. "Stishovite, $SiO_2$, a Very High Pressure New Mineral from Meteor Crater, Arizona." *Jour. Geophys. Res.*, vol. 67, pp. 419–421.

Chao, E. C. T., O. B. James, J. A. Minkin, J. A. Boreman, E. D. Jackson, and C. B. Raleigh. 1970. "Petrology of Unshocked Crystalline Rocks and Evidence of Impact Metamorphism in Apollo 11 Returned Lunar Sample." *Proc. of the Apollo 11 Lunar Science Conference*, vol. 1, pp. 287–314.

Chao, E. C. T., E. M. Shoemaker, and B. M. Madsen. 1960. "First Natural Occurrence of Coesite." *Science*, vol. 132, pp. 220–222.

Chladni, E. F. F. 1794. *Über den Ursprung der von Pallas gefundenen und anderer ihr ahnlicher Eisenmassen, und uber einige damit in Verbindung stehende Naturerscheinungen.* Hartnoch, Riga.

Chorley, R. J., R. P. Beckinsale, and A. J. Dunn. 1973. *The History of the Study of Landforms.* Vol. 2. *The Life and Work of William Morris Davis.* Methuen, London.

Christie, J. M., D. T. Griggs, and N. L. Carter. 1964. "Experimental Evidence of Basal Slip in Quartz." *Jour. Geol.*, vol. 72, pp. 734–756.

Christie, J. M., H. C. Heard, and P. N. LaMori. 1964. "Experimental Deformation of Quartz Single Crystals at 27 to 30 Kilobars Confining Pressure and 24°C." *Am. Jour. Sci.*, vol. 262, pp. 26–55.

Clayton, P. A., and L. J. Spencer. 1934. "Silica-glass from the Libyan Desert." *Mineralog. Mag.*, vol. 23, pp. 501–508.

Coes, L., Jr. 1953. "A New Dense Crystalline Silica." *Science*, vol. 118, pp. 131–132.

Cohen, A. J., T. E. Bunch, and A. M. Reid. 1961. "Coesite Discoveries Establish Cryptovolcanics as Fossil Meteorite Craters." *Science*, vol. 134, pp. 1624–1625.

Coleman, A. P. 1905. "The Sudbury Nickel Region." *Ann. Rept. Ontario Bur. Mines*, 1904, vol. 14, part 3.

———. 1907. "The Sudbury Laccolithic Sheet." *Jour. Geol.*, vol. 15, pp. 759–782.

Coleman, A. P., E. S. Moore, and T. L. Walker. 1929. "The Sudbury Nickel Intrusive." *Contrib. to Canadian Mineralogy.* Univ. of Toronto Press.

Collett, J. 1883. "Geological Survey of Newton County." Indiana Dept. Geol. and Nat. Hist., *12th Ann. Rept.*, pp. 58–59.

Collins, W. H. 1934. "Life History of the Sudbury Nickel Irruptive; I. Petrogenesis." *Trans. Roy. Soc. Canada*, 3rd series, sec. 4, vol. 28, pp. 123–177.

———. 1936. "The Life History of the Sudbury Nickel Irruptive; III. Environment." *Trans. Roy. Soc. Canada*, 3rd series, sec. 4, vol. 30, pp. 29–53.

———. 1937. "The Life History of the Sudbury Nickel Irruptive; IV. Mineralization." *Trans. Roy. Soc. Canada*, 3rd series, sec. 4, vol. 31, pp. 15–43.

Collins, W. H., and E. D. Kindle. 1935. "The Life History of the Sudbury Nickel Irruptive; II. Intrusion and Deformation." *Trans. Roy. Soc. Canada*, 3rd series, sec. 4, vol. 29, pp. 27–47.

Cooke, H. C. 1946. "Problems of Sudbury Geology." *Geol. Survey Bull. No. 3*, Dept. Mines and Resources, Canada.

Cotton, C. A. 1944. *Volcanoes as Landscape Forms.* Whitcombe & Tombs, London.

Cousins, C. A. 1959. "The Structure of the Mafic Portion of the Bushveld Igneous Complex." *Geol. Soc. South Africa Trans.*, vol. 62, pp. 179–189.

Cumings, E. R. 1922. *Nomenclature and Description of the Geological Formations of Indiana.* Indiana Dept. of Conservation, Pubn. no. 21.

Cumings, E. R., and R. R. Shrock. 1927. "Silurian Coral Reefs of Northern Indiana and Their Associated Strata." *Proc. Ind. Acad. Sci.*, vol. 36, pp. 71–85.

———. 1928. "Niagaran Coral Reefs of Indiana and Adjacent States and Their Stratigraphic Relations." *Bull. Geol. Soc. Am.*, vol. 39, pp. 579–620.

Cummings, D. 1965. "Kink-bands: Shock Deformation of Biotite Resulting from a Nuclear Explosion." *Science*, vol. 148, pp. 950–952.

Currie, K. L. 1964a. "Rim Structure of the New Quebec Crater, Canada." *Nature* (January 25), p. 385.

———. 1964b. "On the Origin of Some 'Recent' Craters on the Canadian Shield." *Meteoritics*, vol. 2, no. 2, pp. 93–110.

———. 1965. "Analogues of Lunar Craters on the Canadian Shield." *Annals N.Y. Acad. Sci.*, vol. 123, pp. 915–940.

Dachille, F. 1962. "Interactions of the Earth with Very Large Meteorites." *Bull. South Carolina Acad. Sci.*, vol. 24, pp. 16–34.

Daly, R. A. 1928. "Bushveld Igneous Complex of the Transvaal." *Bull. Geol. Soc. Am.*, vol. 39, pp. 703–768.

———. 1947. "The Vredefort Ring-structure of South Africa." *Jour. Geol.*, vol. 55, pp. 125–145.

Daly, R. A., and G. A. F. Molengraaff. 1924. "Structural Relations of the Bushveld Complex, Transvaal." *Jour. Geol*, vol. 32, pp. 1–35.

Dampier, W. C. 1971. *A History of Science and Its Relations with Philosophy and Religion* (1929). Reprint. Cambridge Univ. Press.

Dana, J. D. 1846. "On the Volcanoes of the Moon." *Am. Jour. Sci.*, vol. 52, pp. 335–355.

Darton, N. H. 1905. "Zuni Salt Lake." *Jour. Geol.*, vol. 13, pp. 185–193.

———. 1945. "Crater Mound, Arizona" (abs.). *Bull. Geol. Soc. Am.*, vol. 56, part 2, p. 1154.

Darwin, C. 1896. *Geological Observations.* D. Appleton & Co., New York. (Originally published in 1844 as *Geological Observations on the Volcanic Islands and Parts of South America Visited During the Voyage of H. M. S. Beagle.*)

Davis, W. M. 1922. "Origin of Coon Butte." In *Biographical Memoirs*, Nat. Acad. Sci. Memoirs, vol. 21, 5th Memoir.

Dawson, J. B. 1967. "A Review of the Geology of Kimberlite." In *Ultramafic and Related Rocks*, edited by P. J. Wyllie. John Wiley & Sons, New York.

DeCarli, P. S., and J. C. Jamieson. 1959. "Formation of an Amorphous Form of Quartz under Shock Conditions." *Jour. Chem. Phys.*, vol. 31, no. 6, pp. 1675–1676.

———. 1961. "Formation of Diamond by Explosive Shock." *Science*, vol. 133, pp. 1821–1822.

DeCarli, P. S., and D. J. Milton. 1965. "Stishovite: Synthesis by Shock Wave." *Science*, vol. 147, pp. 144–145.

Demler, M. C. 1961. "The Role of Very High Pressure in the Air Force Materials Program." In *Progress in Very High Pressure Research*, F. P. Bundy, W. R. Hibbard, and H. M. Strong, eds., pp. xv–xix. John Wiley & Sons, New York.

Dence, M. R. 1964. "A Comparative Structural and Petrographic Study of Probable Canadian Meteorite Craters." *Meteoritics*, vol. 2, no. 3, pp. 249–270.

———. 1965. "Extraterrestrial Origin of Canadian Craters." *Annals N.Y. Acad. Sci.*, vol. 123, pp. 941–969.

———. 1968. "Shock Zoning at Canadian Craters: Petrography and Structural Implications." In *Shock Metamorphism of Natural Materials*, B. M. French and N. M. Short, eds., pp. 169–184. Mono Book Corp., Baltimore.

———. 1971. "Impact Melts." *Jour. Geophys. Res.*, vol. 76, pp. 5552–5565.

Dence, M. R., J. A. V. Douglas, A. G. Plant, and R. J. Traill. 1970. "Petrology, Mineralogy, and Deformation of Apollo 11 Samples. *Proc. of the Apollo 11 Lunar Science Conference*, vol. 1, pp. 315–340.

Dence, M. R., and J. Popelar. 1972. "Evidence for an Impact Origin for Lake Wanapitei, Ontario." Geol. Assoc. of Canada, *Spec. Paper No. 10*, pp. 117–124.

Dence, M. R., P. B. Robertson, and R. L. Wirthlin. 1974. "Coesite from the Lake Wanapitei Crater, Ontario." *Earth and Planetary Science Letters*, 22, pp. 118–122.

Dietz, R. S. 1946a. "The Meteoritic Impact Origin of the Moon's Surface Features. *Jour. Geol.*, vol. 54, pp. 359–375.

———. 1946b. "Geological Structures Possibly Related to Lunar Craters." *Pop. Astron.*, vol. 54, pp. 465–467.

———. 1947. "Meteorite Impact Suggested by Orientation of Shatter Cones at the Kentland, Indiana, Disturbance." *Science*, vol. 105, pp. 42–43.

———. 1959. "Shatter Cones in Cryptoexplosion Structures (Meteorite Impact?)" *Jour. Geol.*, vol. 67, pp. 496–505.

———. 1960. "Meteorite Impact Suggested by Shatter Cones in Rock." *Science*, vol. 131, pp. 1781–1784.

———. 1961. "Vredefort Ring Structure: Meteorite Impact Scar?" *Jour. Geol.*, vol. 69, pp. 499–516.

———. 1962. "Vredefort Ring-Bushveld Complex Impact Event and Lunar Maria" (abs.). *Geol. Soc. Am. Spec. Paper No. 73*, p. 35.

———. 1963a. "Cryptoexplosion Structures: A Discussion." *Am. Jour. Sci.*, vol. 261, pp. 650–664.

———. 1963b. "Astroblemes: Ancient Meteorite-Impact Structures on the Earth." In *The Moon, Meteorites, and Comets*, B. M. Middlehurst and G. P. Kuiper, eds., pp. 285–300. Univ. of Chicago Press.

———. 1964. "Sudbury Structure as an Astrobleme." *Jour. Geol.*, vol. 72, pp. 412–434.

———. 1972. "Sudbury Astrobleme, Splash Emplaced Sub-layer and Possible Cosmogenic Ores." Geol. Assoc. of Canada, *Special Paper No. 10*, pp. 29–40.

———. 1977. "Elgygytgyn Crater, Siberia: Probable Source of Australasian Tektite Field." *Meteoritics*, vol. 12, pp. 145–157.

Dietz, R. S., and J. P. Barringer. 1973. "Hudson Bay Arc as an Astrobleme: A Negative Search." *Meteoritics*, vol. 8, pp. 28–29.

Dietz, R. S., and L. W. Butler. 1964. "Shatter-cone Orientation at Sudbury, Canada." *Nature* (Oct. 17), pp. 280–281.

Dietz, R. S., R. Fudali, and W. Cassidy. 1969. "Richat and Semsiyat Domes (Mauritania): Not Astroblemes." *Bull. Geol. Soc. Am.*, vol. 80, pp. 1367–1372.

Downes, P. G. 1943. *Sleeping Island.* Coward McCann, New York.

du Toit, A. L. 1954. *Geology of South Africa* (3rd edition). Oliver and Boyd, Edinburgh and London.

Dutton, C. E. 1885. "Mount Taylor and the Zuni Plateau." *U.S.G.S. 6th Ann. Rept.*, pp. 105–198.

Dyer, R. E. H. 1921. *Raiders of the Sarhad.* H. F. and G. Witherby, London.

El Goresy, A. 1965. "Baddeleyite and Its Significance in Impact Glasses." *Jour. Geophys. Res.*, vol. 70, pp. 3453–3456.

Elrod, M. N., and A. C. Benedict. 1891. "Geology of Wabash County." Indiana Dept. Geol. and Nat. Resources, *17th Ann. Rept.*, pp. 200–227.

Engelhardt, W.v., and F. Hörz. 1965. "Riesgläser und Moldavite." *Geochim. et Cosmochim. Acta*, vol. 29, pp. 609–620.

Engeln, O. D. von. 1932. "The Ubehebe Craters and Explosion Breccia in Death Valley, California." *Jour. Geol.*, vol. 40, pp. 726–734.

Fairbairn, H. W. 1939. "Correlation of Quartz Deformation with Its Crystal Structure." *Am. Mineralogist,* vol. 24, pp. 351–368.

Fairbairn, H. W., P. M. Hurley, and W. H. Pinson. 1960. "Mineral and Rock Ages at Sudbury-Blind River. Ontario." *Proc. Geol. Assoc. of Canada,* vol. 12, pp. 41–64.

Fairbairn, H. W., and G. M. Robson. 1942. "Breccia at Sudbury, Ontario." *Jour. Geol.,* vol. 50, pp. 1–33.

Fairchild, H. L. 1907. "Origin of Meteor Crater (Coon Butte), Arizona." *Bull. Geol. Soc. Am.,* vol. 18, pp. 493–504.

Farrington, O. C. 1915. *Meteorites.* Chicago (published by the author).

Faul, H. 1966. "Tektites Are Terrestrial." *Science,* vol. 152, pp. 1341–1345.

Fisher, C. 1943. *The Story of the Moon.* Am. Mus. Nat. Hist. Sci. Series, Doubleday, Doran & Co., Inc., Garden City, N.Y.

Fleischer, R. L., and P. B. Price. 1964. "Fission Track Evidence for the Simultaneous Origin of Tektites and Other Natural Glasses." *Geochim. et Cosmochim. Acta,* vol. 28, pp. 755–760.

Fleischer, R. L., P. B. Price, and R. M. Walker. 1965. "On the Simultaneous Origin of Tektites and Other Natural Glasses." *Geochim. et Cosmochim. Acta,* vol. 29, pp. 161–166.

Foerste, A. F. 1921. "Notes on Arctic Ordovician and Silurian Cephalopods. *Jour. of the Scientific Labs.,* Dennison University, vol. 19, pp. 277–278.

Foote, A. E. 1891. "A New Locality for Meteoric Iron, with a Preliminary Notice of the Discovery of Diamonds in the Iron." *Am. Jour. Sci.,* 3rd series, vol. 42, pp. 413–417.

Ford, F. J. 1972. "A Possible Impact Crater Associated with Darwin Glass." *Earth and Planetary Science Letters,* vol. 16, pp. 228–230.

Fourcroy, A. F. 1803. "Memoir on the Stones Which Have Fallen from the Atmosphere. . . ." *Phil. Mag. (Tilloch's),* vol. 16, pp. 299–305.

Fredriksson, K., A. Dube, and D. Milton. 1973. "Microbreccias, Impact Glasses and Spherules from Lonar Lake Crater, India: Lunar Analogs." *Meteoritics,* vol. 8, p. 34.

Fredriksson, K., D. Milton, A. Dube, and M. S. Balasundaram. 1973. "The Lonar Meteorite Crater, India." *Meteoritics,* vol. 8, p. 35.

French, B. M. 1967. "Sudbury Structure, Ontario: Some Petrographic Evidence for an Origin by Meteorite Impact." *Science,* vol. 156, pp. 1094–1098.

————. 1968. "Sudbury Structure, Ontario: Some Petrographic Evidence for an Origin by Meteorite Impact." In *Shock Metamorphism of Natural Materials,* edited by B. M. French and N. M. Short, pp. 383–412. Mono Book Corp., Baltimore.

————. 1970. "Possible Relations Between Meteorite Impact and Petrogenesis, as Indicated by the Sudbury Structure, Ontario, Canada." *Bull. Volcanol.,* vol. 34, pp. 466–517.

French, B. M., and R. B. Hargraves. 1971. "Bushveld Igneous Complex, South Africa: Absence of Shock-Metamorphic Effects in a Preliminary Search." *Jour. Geol.,* vol. 79, pp. 616–620.

French, B. M., J. B. Hartung, N. M. Short, and R. S. Dietz. 1970. "Tenoumer Crater, Mauritania: Age and Petrologic Evidence for Origin by Meteorite Impact." *Jour. Geophys. Res.*, vol. 75, pp. 4396–4406.

French, B. M., and N. M. Short, eds. 1968. *Shock Metamorphism of Natural Materials.* Mono Book Corp., Baltimore.

Fudali, R.F., and W. A. Cassidy. 1972. "Gravity Reconnaissance at Three Mauritanian Craters of Explosive Origin." *Meteoritics*, vol. 7, pp. 51–70.

Fudali, R. F., and R. J. Ford. 1978. "Darwin 'Crater,' Tasmania: A Progress Report.' *Meteoritics*, vol. 13, p. 471.

————. 1979. "Darwin Glass and Darwin Crater: A Progress Report." *Meteoritics*, vol. 14, pp. 283–296.

Fudali, R. F., D. P. Gold, and J. J. Gurney. 1973. "The Pretoria Salt Pan: Astrobleme or Cryptovolcano?" *Jour. Geol.*, vol. 81, pp. 495–507.

Gasteyer, C. E. 1970. "Forest Ray Moulton." *Nat. Acad. Sci. Biograph. Memoirs*, vol. 41, pp. 341–355.

Gault, D. E., J. E. Guest, J. B. Murray, D. Dzurisin, and M. Malin. 1975. "Some Comparisons of Impact Craters on Mercury and on the Moon." *Jour. Geophys. Res.*, vol. 80, pp. 2444–2460.

Gentner, W., H. J. Lippolt, and O. Müller. 1964. "Das Kalium-Argon-Alter des Bosumtwi-Kraters in Ghana und die chemische Beschaffenheit seiner Gläser." *Zeitschrift für Naturforschung*, vol. 19*a*, pp. 150–153.

Gentner, W., H. J. Lippolt, and O. A. Schaeffer. 1963. "Argonbestimmungen an Kaliummineralien — XI. Die Kalium-Argon-Alter der Gläser des Nördlinger Rieses und der bohmisch-mährischen Tektite." *Geochim. et Cosmochim. Acta*, vol. 27, pp. 191–200.

Gifford, A. C. 1924. "The Mountains of the Moon." *New Zealand Jour. Sci. and Technol.*, vol. 7, pp. 129–142.

————. 1930. "The Origin of the Surface Features of the Moon." *Scientia*, vol. 48, pp. 69–80.

Gilbert, G. K. 1877. *Report on the Geology of the Henry Mountains.* Reprint. Arno Press, Inc., 1978.

————. 1886. "Inculcation of Scientific Method by Example with an Illustration Drawn from the Quaternary Geology of Utah." *Am. Jour. Sci.*, vol. 31, pp. 284–299.

————. 1892. *U.S.G.S. 13th Ann. Rept.*, p. 98.

————. 1893*a*. (given orally in 1892). "The Moon's Face: A Study of the Origin of Its Features." *Bull. Phil. Soc. of Washington*, vol. 12, pp. 241–292.

————. 1893*b. U.S.G.S. 14th Ann. Rept.*, p. 187.

————. 1896. "The Origin of Hypotheses, Illustrated by the Discussion of a Topographic Problem." *Science*, new series, 3, pp. 1–13.

Gilbert, G. K., and M. Baker. 1891. "A Meteoric Crater." *Astron. Soc. of the Pacific*, no. 21, p. 37.

Gilbert, G. K., A. R. Marvine, E. E. Howell, and J. J. Stevens. 1875. *Reports on Geographical and Geological Explorations and Sur-*

*veys West of the 100th Meridian.* Vol. 3. *Geology.* Government Printing Office, Washington, D.C.

Gilvarry, J. J., and J. E. Hill. 1956. "The Impact of Large Meteorites." *Astrophys. Jour.,* vol. 24, pp. 610–622.

Goles, G. G., R. A. Fish, and E. Anders. 1960. "The Record in the Meteorites — I: The Former Environment of Stone Meteorites as Deduced from $K^{40}-Ar^{40}$ Ages." *Geochim. et Cosmochim. Acta,* vol. 18, pp. 177–195.

Gorby, S. S. 1886. "The Wabash Arch." Indiana Dept. Geol. and Nat. Resources, *15th Ann. Rept.,* pp. 228–241.

Grant, K., and G. F. Dodwell. 1931. "The Karoonda (S.A.) Meteorite of Nov. 25, 1930." *Nature,* vol. 127, pp. 402–403.

Green, J. 1965. "Hookes and Spurrs in Selenology." *Annals N.Y. Acad. Sci.,* vol. 123, pp. 373–402.

Greene, G. K. 1906. *Contributions to Indiana Paleontology,* vol. 2, part 1, pp. 11–17.

Griggs, D. 1936. "Deformation of Rocks Under High Confining Pressures." *Jour. Geol.,* vol. 44, pp. 541–577.

————. 1938. "Deformation of Single Calcite Crystals Under High Confining Pressures." *Am. Mineralogist,* vol. 23, pp. 28–33.

Griggs, D., and J. F. Bell. 1938. "Experiments Bearing on the Orientation of Quartz in Deformed Rocks." *Bull. Geol. Soc. Am.,* vol. 49, pp. 1723–1746.

Griggs, D., and W. B. Miller. 1951. "Deformation of Yule Marble: Pt. I — Compression and Extension Experiments on Dry Yule Marble at 10,000 Atmospheres Confining Pressure, Room Temperature." *Bull. Geol. Soc. Am.,* vol. 62, pp. 853–862.

Griggs, D., F. J. Turner, I. Borg, and J. Sosoka. 1951. "Deformation of Yule Marble: Part IV—Effects at 150°C." *Bull. Geol. Soc. Am.,* vol. 62, pp. 1385–1406.

————. 1953. "Deformation of Yule Marble: Part V—Effects at 300°C." *Bull. Geol. Soc. Am.,* vol. 64, pp. 1327–1342.

Gunther, R. T. 1939. "Dr. William Thomson, F. R. S., a Forgotten English Mineralogist, 1761–c. 1806." *Nature,* vol. 143, pp. 667–668.

Guppy, D. J., and R. S. Matheson. 1950. "Wolf Creek Meteorite Crater, Western Australia." *Jour. Geol.,* vol. 58, pp. 30–36.

Hack, J. T. 1942. "Sedimentation and Volcanism in the Hopi Buttes." *Bull. Geol. Soc. Am.,* vol. 53, pp. 335–372.

Hager, D. 1926. "Meteor Crater." Letter in *Eng. and Min. Journal-Press,* vol. 121, p. 374.

————. 1953. "Crater Mound (Meteor Crater), Arizona, a Geologic Feature." *Bull. Am. Assoc. Petrol. Geol.,* vol. 37, pp. 821–857.

Hall, A. L. 1906. "Geological Notes on the Bushveld Tin Fields and the Surrounding Area." *Trans. Geol. Soc. South Africa,* vol. 8, pp. 47–55.

————. 1914. "The Bushveld as a Metamorphic Province." Presidential address, *Proc. Geol. Soc. South Africa,* vol. 17, pp. 22–37.

————. 1932. "The Bushveld Igneous Complex of the Central Transvaal." *Geol. Survey Mem. No. 28,* Government Printer, Pretoria.

Hall, A. L., and G. A. F. Molengraaff. 1925. *The Vredefort Mountain Land in the Southern Transvaal and the Northern Orange Free State.* Shaler Memorial Series, Amsterdam.

Halley, E. 1714. "An Account of Several Extraordinary Meteors or Lights in the Sky." *Phil. Trans. Roy. Soc.,* No. 341, 4.

———. 1719. "An Account of the Extraordinary Meteor Seen all over England on the 19th of March 1718/19." *Phil. Trans. Roy. Soc.,* No. 360, 7.

Halliday, I., and A. A. Griffin, 1963. "Evidence in Support of a Meteoritic Origin for West Hawk Lake, Manitoba, Canada." *Jour. Geophys. Res.,* vol. 68, no. 18.

———. 1964. "Application of the Scientific Method to Problems of Crater Recognition." *Meteoritics,* vol. 2, pp. 79–84.

Hamilton, W. 1961. "Form of the Sudbury Lopolith." *Can. Mineralog.,* vol. 6, pp. 437–447.

———. 1970. "Bushveld Complex—Product of Impacts?" *Geol. Soc. South Africa Spec. Pub. 1,* pp. 367–374.

Handin, J. W., and D. Griggs, 1951. "Deformation of Yule Marble: Part II—Predicted Fabric Changes." *Bull. Geol. Soc. Am.,* vol. 62, pp. 863–885.

Hardy, C. T. 1953. "Structural Dissimilarity of Meteor Crater and Odessa Meteorite Craters." *Bull. Am. Assoc. Petrol. Geol.,* vol. 37, p. 2580.

———. 1954. "Major Craters Attributed to Meteoritic Impact." *Bull. Am. Assoc. Petrol. Geol.,* vol. 38, pp. 917–923.

Hargraves, R. B. 1961. "Shatter Cones in the Rocks of the Vredefort Ring." *Geol. Soc. South Africa Trans.,* vol. 64, pp. 147–161.

Harrison, J. M. 1954. "Ungava (Chubb) Crater and Glaciation." *Roy. Astron. Soc. Can. Jour.,* vol. 48, pp. 16–20.

Harrison, T. S. 1927. "Colorado-Utah Salt Domes." *Bull. Am. Assoc. Petrol. Geol.,* vol. 11, pp. 111–133.

Harwood, H. F. 1916. "Some Fresh Analyses of the Granite and Pseudotachylite of Parijs." *Quart. Jour. Geol. Soc. South Africa,* vol. 31, pp. 217–219.

Hawley, J. E. 1962. "The Sudbury Ores: Their Mineralogy and Origin." *Can. Mineralog.,* vol. 7, pp. 1–203.

Healy, P. W., L. LaPaz, and F. C. Leonard. 1953. "On the Identification of Terrestrial Meteorite Craters." *Roy. Astron. Soc. Can. Jour.,* vol. 47, pp. 160–161.

Hendriks, H. E. 1954. *Geology of the Steelville Quadrangle, Mo.* State of Missouri Business & Administration, Div. of Geol. Surv. and Water Resources, vol. 36, 2nd series.

Herrick, C. R. 1900. "Report on a Geological Reconnaissance in West Socorro and Valencia Counties (New Mexico)." *Am. Geol.,* vol. 25, pp. 331–346.

Hoffleit, D. 1952. "Murgab Meteorite Craters." *Sky and Telescope* (November), p. 8.

Hörz, F. 1965. "Untersuchungen an Riesgläsern." *Beiträge zur Min. und Petrog.,* vol. 11, pp. 621–661.

Howard, E. C. 1802. "Experiments and Observations on Certain

Stony and Metalline Substances, Which at Different Times Are Said To Have Fallen on the Earth, etc."*Phil. Trans. Roy. Soc.*, pp. 168–212.

Hoyt, W. G. 1983. "Meteor Crater: Historical Note on Nomenclature." *Meteoritics*, vol. 18, no. 2, pp. 159–163.

Humphrey, W. A. 1908. "The Geology of the Neighbourhood of Rustenburg."*Ann. Rept. Geol. Surv.*, Transvaal, Pretoria, pp.81–90.

———. 1909. "On a Portion of the Bushveld Bordering the Crocodile River and Including the Rooiberg Tin-field." *Ann. Rept. Geol. Surv.*, Transvaal, Pretoria, for 1908, pp.105–122.

Innes, M. J. S. 1957. "A Possible Meteorite Crater at Deep Bay. Saskatchewan." *Roy. Astron. Soc. Can. Jour.*, vol. 51, no. 4, pp. 235–240.

———. 1964. "Recent Advances in Meteorite Crater Research at the Dominion Observatory, Ottawa, Canada." *Meteoritics*, vol. 2, pp. 219–241.

Innes, M. J. S., W. H. Pearson, and J. W. Geuer. 1964. "The Deep Bay Crater." *Pubn. of the Dominion Observatory, Ottawa, Canada*, vol. 31, no. 2, pp.19–52.

Ives, H. E. 1919. "Some Large-Scale Experiments Imitating the Craters of the Moon." *Astrophys. Jour.*, vol. 50, pp. 245–250.

Jakosky, J. J. 1932. "Geophysical Methods Locate Meteorite." *Eng. & Min. Jour.*, vol. 133, no. 7, pp. 392–393.

Janssen, C. L. 1951. "The Meteor Craters in Hérault, France." *Roy. Astron. Soc. Can. Jour.*, vol. 45, pp. 190–198.

Jorissen, E. 1905. "Notes on Some Intrusive Granites in the Transvaal, the Orange River Colony, and in Swaziland." *Geol. Soc. South Africa Trans.*, vol. 7, pp. 151–160.

Joule, J. P. 1848. "On Shooting Stars." *Phil. Mag.*, series 2, pp. 349–351.

Karpoff, R. 1953. "The Meteorite Crater of Talemzane in Southern Algeria (CN=∓0041,333). *Meteoritics*, vol. 1, pp. 31–38.

Keefe, J. A. C., S. H. Ward, and J. T. Wilson. 1957. "Report on Survey of Vertical Magnetic Intensity Near Chubb Crater, July–August, 1951." *Roy. Astron. Soc. Can. Jour.*, vol. 51, pp. 148–154.

Kelley, V. C. 1956. "Influence of Regional Structure and Tectonic History upon the Origin and Distribution of Uranium and Thorium on the Colorado Plateau." *U.S.G.S. Prof. Paper No. 300.*

Kerker, M. 1960. "Sadi Carnot and the Steam Engine Engineers." *Isis*, vol. 51, pp. 257–270.

Kindle, E. M. 1903. "The Stratigraphy and Paleontology of the Niagara of Northern Indiana." Indiana Dept. of Geol. and Nat. Resources, *28th Ann. Rept.*, pp. 401–428.

King, P. B. 1930. "The Geology of the Glass Mountains, Texas. Part I. Descriptive Geology: Sierra Madera." *Univ. of Texas Bull. 3038*, pp. 123–125.

Klaproth, M. H. 1803. "Des Masses pierreuses et métalliques tombées de l'atmosphere." *Mem. de l'Acad. Royale des Sciences et Belles Lettres*, pp. 37–66, Berlin.

Knight, C. W. 1916. "Origin of Sudbury Nickel-Copper Deposits." *Eng. and Min. Jour.*, vol. 101, pp. 811–812.

Knopf, E. B. 1933. "Petrotectonics." *Am. Jour. Sci.*, 5th series, vol. 25, no. 150, pp. 433–470.

———. 1946. "Study of Experimentally Deformed Rocks." *Science*, vol. 103, pp. 99–103.

Kopal, Z. 1971. *A New Photographic Atlas of the Moon.* Taplinger Publishing Co., New York.

Krinov, E. L. 1956. "The Siberian Meteorite Fall of February, 1947." *Sky and Telescope* (May), pp. 300–301.

———. 1960a. *Principles of Meteoritics.* Translated from the Russian by Irene Vidziunas. Translation edited by Harrison Brown. Pergamon Press, London.

———. 1960b. "The Tunguska Meteorite." *Internat. Geol. Review*, vol. 2, pp. 8–9.

———. 1963. "The Tunguska and Sikhote-Alin Meteorites." In *The Moon, Meteorites, and Comets*, edited by B. M. Middlehurst and G. P. Kuiper, pp. 208–234. Univ. of Chicago Press.

———. 1966. *Giant Meteorites* (first English edition). Translated from the Russian by J. S. Romankiewicz. Translation edited by M. M. Beynon. Pergamon Press, London.

Kuiper, G. P. 1954. "On the Origin of the Lunar Surface Features." *Proc. Nat. Acad. Sci.*, vol. 40, no. 12, pp. 1096–1112.

Kulik, L. 1928. "Auffindung des tungusischen Riesenmeteors vom 30 Juni 1908." *Petermann's Geografische Mitteilungen*, vol. 74, pp. 338–341.

———. 1937. "The Question of the Meteorite of June 30, 1908, in Central Siberia." *Pop. Astron.*, vol. 45, pp. 559–562.

Kunz, G. F. 1890. "On Five New American Meteorites." *Am. Jour. Sci.*, 3rd series, vol. 40, pp. 312–323.

Kurat, G., and W. Richter. 1969. "On the 'Pumice' from the Köfels Crater, Austria." *Meteoritics*, vol. 4, p. 192.

Kynaston, H. 1910a. "The Red Granite of the Transvaal Bushveld and Its Relation to Ore Deposits." *Proc. Geol. Soc. South Africa*, vol. 12 (1909), pp. 21–30.

———. 1910b. "On a Portion of the Waterberg District West and Northwest of Warmbaths." *Ann Rept. Geol. Surv.*, Transvaal, for 1909, pp. 23–38.

Lacroix, A. 1924. "On a New Type of Meteoric Iron Found in the Desert of Adrar in Mauritania." *Meteoritics*, 1, pp. 192–196, (translated from the French of A. Lacroix, *Comptes Rendus*, vol. 179, pp. 303–313, 1924).

Lafond, E. C., and R. S. Dietz. 1964. "Lonar Lake, India, a Meteorite Crater?" *Meteoritics*, vol. 2, pp. 111–116.

LaPaz, L. 1953. "The Discovery and Interpretation of Nickel-Iron Granules Associated with Meteorite Craters." *Roy. Astron. Soc. Can. Jour.*, vol. 47, pp. 191–194.

LaPaz, L. and J. 1954. "The Adrar (Chinguetti), Mauritania, French West Africa, Meteorite (CN=0127,202)." *Meteoritics*, vol. 1, pp. 187–196.

LeGrand, H. E. 1953. "Streamlining of the Carolina Bays." *Jour. Geol.*, vol. 61, pp. 263–274.

Leonard, F. C. 1946. "Authenticated Meteorite Craters of the World: A Catalog of Provisional Coordinate Numbers for the Meteoritic Falls of the World." *Univ. of New Mexico Pubns. in Meteoritics*, no. 1, p. 54. Univ. of New Mexico Press.

Lipschutz, M. E. 1962. "Diamonds in the Dyalpur Meteorite," *Science*, vol. 138, pp. 1266–1267.

———. 1964. "Origin of Diamonds in the Ureilites." *Science*, vol. 143, pp. 1431–1434.

Lipschutz, M. E., and E. Anders. 1961. "The Record in the Meteorites —IV: Origin of Diamonds in Iron Meteorites." *Geochim. et Cosmochim. Acta*, vol. 24, pp. 83–105.

Littler, J., J. J. Fahey, R. S. Dietz, and E. C. T. Chao. 1961. "Coesite from the Lake Bosumtwi Crater, Ashanti, Ghana" (abs.). *Geol. Soc. Am. Special Paper No. 68*, p. 218.

Lockyer, J. N. 1890. *The Meteoritic Hypothesis.* Macmillan & Co., London.

Lovering, J.F. 1957. "Pressures and Temperatures Within a Typical Parent Meteorite Body." *Geochim. et Cosmochim. Acta*, vol. 12, pp. 253–261.

MacDonald, G. J. F. 1956. "Quartz-Coesite Stability Relations at High Temperatures and Pressures." *Am. Jour. Sci.*, vol. 254, pp. 713–721.

MacDonald, T. L. 1931. "The Distribution of Lunar Altitudes." *British Astron. Assoc.*, vol. 41, pp. 172–183 and 228–239.

MacLaren, M. 1931. "Lake Bosumtwi, Ashanti." *Geograph. Jour.*, vol. 78, pp. 270–276.

Madigan, C. T. 1937. "The Boxhole Crater and the Huckitta Meteorite (Central Australia)." *Roy. Soc. South Australia Trans.*, vol. 61, pp. 187–190.

———. 1940. "The Boxhole Meteoritic Iron, Central Australia." *Mineralog. Mag.*, vol. 25, pp. 481–486.

Magie, W. F. 1910. "Physical Notes on Meteor Crater, Arizona." *Proc. Am. Phil. Soc.*, vol. 49, pp. 41–48.

Manton, W. I. 1965. "The Orientation and Origin of Shatter Cones in the Vredefort Ring." *Annals of N.Y. Acad. Sci.*, vol. 123, pp. 1017–1049.

Maree, B. D. 1944. "The Vredefort Structure as Revealed by a Gravimetric Survey." *Geol. Soc. South Africa Trans.*, vol. 47, pp. 183–196.

Martini, J. E. J. 1978. "Coesite and Stishovite in the Vredefort Dome, South Africa." *Nature*, vol. 272, pp. 715–717.

Marvin, U. 1986. "Meteoritics, the Moon, and the History of Geology." *Jour. of Geological Education*, vol. 34, pp. 140–165. (There is no reference to this work in text, as it appeared when this volume was already in proof.)

Masaitis, V. L., S. I. Futergendler, and M. A. Gnevushev. 1972. "Diamonds in Impactites of the Popigai Meteorite Crater" (in Russian).

*Zapiski Vsesoyuznogo Mineralogicheskogo Obshchestva*, no. 1 (Reports of the All-Union Mineralogical Society), pp. 108–112.

Masaitis, V. L., M. V. Mikhailov, and T. V. Selivanovskaya. 1971. "Popigai Basin—An Explosion Meteorite Crater." Translated by T. K. Bodine from *Doklady Akademii Nauk SSSR*, vol. 197, no. 6, 1971. *Meteoritics*, vol. 7, 1972, pp. 39–46.

Mason, B. 1962. *Meteorites*. John Wiley & Sons, New York.

McCall, G. J. H. 1964. "Are Cryptovolcanic Structures Due to Meteoric Impact?" *Nature*, vol. 201, pp. 251–254.

McIntyre, D. B. 1962. "Impact Metamorphism at Clearwater Lake, Quebec" (abs.). *Jour. Geophys. Res.*, vol. 67, part 2, p. 1647.

McKnight, E. T. 1940. "Geology of Area Between Green and Colorado Rivers, Grand and San Juan Counties, Utah." *U.S.G.S. Bull. 908*, pp. 124–128.

Meen, V. B. 1950. "Chubb Crater, Ungava, Quebec." *Roy. Astron. Soc. Can. Jour.*, vol. 44, pp. 169–180.

———. 1951a. "Chubb Crater, Ungava, Quebec." *Geol. Assoc. Can. Proc.*, vol. 4, pp. 49–59.

———. 1951b. "The Canadian Meteor Crater." *Sci. Am.*, vol. 184, pp. 64–69.

———. 1952. "Solving the Riddle of Chubb Crater." *Nat. Geog. Mag.*, vol. 101, no. 1.

———. 1957. "Merewether Crater: A Possible Meteor Crater." *Proc. Geol. Assoc. Can.*, vol. 9, pp. 49–67.

Mendenhall, W. C. 1920. "Memorial: Grove Karl Gilbert." *Bull. Geol. Soc. Am.*, vol. 31, pp. 26–64.

Merrill, G.P. 1907. "On a Peculiar Form of Metamorphism in Siliceous Sandstone." *Proc. U.S. Nat. Mus.*, vol. 32, pp. 547–551.

———. 1908. "The Meteor Crater of Canyon Diablo, Arizona: Its History, Origin, and Associated Meteoric Irons." *Smithsonian Inst. Misc. Coll. 1.*, pp. 461–498.

———. 1922. "Meteoritic Iron from Texas." *Am. Jour. Sci.*, series 5, vol. 3, pp. 335–337.

———. 1928. "The Siberian Meteorite." *Science*, vol. 67, pp. 489–490.

Middlehurst, B. M., and G. P. Kuiper (eds.). 1963. *The Moon, Meteorites, and Comets*. Univ. of Chicago Press.

Milledge, H. J. 1961. "Coesite as an Inclusion in G. E. C. Synthetic Diamonds." *Nature*, 190, no. 4782.

Miller, A. H., and M. J. S. Innes. 1955. "Gravity in the Sudbury Basin and Vicinity." *Pubn. of the Dominion Observatory, Ottawa, Canada*, vol. 18, no. 2.

Miller, A. M. 1923. "Meteorites." *Sci. Monthly*, vol. 17, pp. 435–448.

Millman, P. M. 1956. "A Profile Study of the New Quebec Crater." *Pubn. of the Dominion Observatory, Ottawa, Canada*, vol. 18, no. 4.

Millman, P. M., B. A. Liberty, J. F. Clark, P. L. Willmore, and M. J. S. Innes. 1960. "The Brent Crater." *Pubn. of the Dominion Observatory, Ottawa, Canada*, vol. 24, no. 1.

Milton, D.J., and P. S. DeCarli. 1963. "Maskelynite: Formation by Explosive Shock." *Science,* vol. 140, pp. 670–671.

Milton, D. J., and C. W. Naesser. 1971. "Evidence for an Impact Origin of the Pretoria Salt Pan, South Africa." *Nature Physical Science,* vol. 229, pp. 211–212.

Molengraaff, G. A. F. 1904 (first publication in French, 1901). *The Geology of the Transvaal,* translated by J. H. Ronaldson. Edinburgh and Johannesburg.

Monnig, O. E., and R. Brown. 1935. "The Odessa, Texas, Meteorite Crater." *Pop. Astron.,* vol. 43, pp. 34–37.

Monod, T. 1955. "The Problem of the Chinguetti (French West Africa) Meteorite (CN=0127,202)." *Meteoritics,* vol. 1, pp. 308–314.

Monod, T., and A. Pourquié. 1951. "Le cratère d'Aouelloul (Adrar, Sahara Occidental)." *Bull. de l'Inst. francaise d'Afrique noire,* vol. 13, pp. 293–304.

Moulton, F. R. 1929a. Report dated Aug. 24, 1929. Mimeograph copy on file in Lowell Observatory Library, Flagstaff, Arizona.

———. 1929b. Letters on the Arizona Meteor Crater, Aug. 28 to Sept. 24, 1929. On file in the Lowell Observatory Library, Flagstaff, Arizona.

———. 1929c. Second Report. Mimeograph copy on file in Lowell Observatory Library, Flagstaff, Arizona.

———. 1931. *Astronomy*: Macmillan Co., New York.

Müller, G., and G. Veyl, 1957. "The Birth of Nilahue, a New Maar Type." *Internat. Geol. Cong. Rept. 20,* sec. 1, pp. 375–396.

Murray, B., M. C. Malin, and R. Greeley. 1981. *Earthlike Planets.* W. H. Freeman and Co., San Francisco.

Nágara, J. J. 1926. *Los Hoyos del Campo del Cielo y el Meteorito.* Pub. No.19, Argentina Republic Instituto Nacional de Geología y Minería, Buenos Aires.

Nel, L. T. 1927. *The Geology of the Country Around Vredefort: An Explanation of the Geological Map.* Spec. Pubn. Geol. Surv. South Africa, Pretoria.

Nicolaysen, L. O., J. W. L. de Villiers, A. J. Burger, and F. W. E. Strelow. 1958. "New Measurements Relating to the Absolute Age of the Transvaal System and of the Bushveld Igneous Complex." *Trans Geol. Soc. South Africa,* vol. 41, pp 137–163.

Nininger, H. H. 1933. *Our Stone-Pelted Planet.* Houghton Mifflin Co., New York.

———. 1936. "Kansas Meteorites Since 1925." *Trans. Kansas Acad. Sci.,* vol. 39, pp. 169–183.

———. 1938. "Reply to Dr. Spencer's paper, 'Meteorites and the Craters of the Moon'." *Pop. Astron.,* vol. 46, pp. 107–109.

———. 1943. "The Moon as a Source of Tektites. I and II." *Sky and Telescope,* vol. 2, no. 4, pp. 12–15, and no. 5, pp. 8–9.

———. 1949. "Oxidation Studies at Barringer Crater." *Am. Phil. Soc. Yearbook,* pp. 126–133.

———. 1952. *Out of the Sky.* Univ. of Denver Press.

———. 1954. "Reply to Dorsey Hager." *Am. Jour. Sci.,* vol. 252, pp. 697–700.

———. 1956. *Arizona's Meteorite Crater.* Am. Meteorite Laboratory, Denver, Colo.

———. 1959. "Another Meteorite Crater Studied." *Science* (Nov.), pp. 1251–1252.

———. 1972. *Find a Falling Star.* Paul S. Ericksson, Inc., New York.

Nininger, H. H., and J. D. Figgins. 1933. "The Excavation of a Meteorite Crater Near Haviland, Kiowa County, Kansas." *Proc. Colo. Mus. Nat. Hist.,* vol. 12, no. 3, pp. 9–16.

Nininger, H. H., and G. I. Huss. 1961. "The Unique Meteorite Crater at Dalgaranga, Western Australia." *Mineralog. Mag.,* vol. 32, pp. 619–639.

Öpik, E. 1916. "Note on the Meteoric Theory of Lunar Craters" (in Russian with French summary). *Bull. Soc. Russe des Amis de l'Etude de L'Univers* (Mirove denie), vol. 5, pp. 125–134.

Orlov, V. L. 1973. *The Mineralogy of the Diamond.* Translation from the Russian by Izdatel'stva Nauka. John Wiley & Sons, New York, 1977.

Paneth, F. A. 1960. "The Discovery and Earliest Reproductions of the Widmanstätten Figures." *Geochim. et Cosmochim. Acta,* vol. 18, pp. 176–182.

Patterson, J. H., A. L. Turkevich, E. J. Franzgrote, T. E. Economou, and K. P. Sowinski, 1970. "Chemical Composition of the Lunar Surface in a Terra Region Near the Crater Tycho." *Science,* vol. 168, pp. 825–828.

Peck, E. L. 1979. *Space Rocks and Buffalo Grass.* Peach Enterprises, Inc., Warren, Mich.

Pecora, W. T. 1960. "Coesite and Space Geology." *Geotimes,* vol. 5, pp. 16–19.

Perkins, J. 1820. "On the Compressibility of Water." *Phil. Trans. Roy. Soc.,* vol. 154–155, pp. 324–329.

Phemister, T. C. 1925. "Igneous Rocks of Sudbury and Their Relation to Ore Deposits." *34th Ann. Rept. of Ontario Dept. Mines,* vol. 34, part 8.

Philby, H. St. J. 1933. *The Empty Quarter.* Constable & Co., London.

Phinney, A. J. 1890. "The Natural Gas Field of Indiana." *U.S.G.S. 11th Ann. Rept.,* part 1, pp. 617–620.

Playfair, J. 1802. *Illustrations of the Huttonian Theory of the Earth.* Reprint. Dover, 1964.

Pliny. 1938. *Natural History I,* translated by H. A. Rackham. Harvard Univ. Press.

Poldervaart, A. 1962. "Notes on the Vredefort Dome." *Geol. Soc. South Africa Trans.,* vol. 65, pp. 231–251.

Pringle, J. 1759. "Several Accounts of the Fiery Meteor Which Appeared on Sunday the 26th of November, 1758. . . ." *Phil. Trans. Roy. Soc.,* XVI, XVII, XVIII.

Prior, G. T., and M. H. Hey. 1953. *Catalogue of Meteorites* (2nd ed., revised and enlarged). British Museum (Natural History), London.

Proctor, R. A. 1873. *The Moon: Her Motion, Aspect, Scenery, and Physical Condition.* D. Appleton & Co., New York.

Prommel, H. W. C., and H. E. Crum. 1927. "Salt Domes of Permian and Pennsylvanian Age in Southeastern Utah and Their Influence on Oil Accumulation." *Bull. Am. Assoc. Petrol. Geol.*, vol. 11, pp. 373–393.

Proust, J. L. 1800. "Account of a Memoir of M. Proust, on Several Interesting Points in Chemistry." *Jour. Nat. Phil. (Nicholson's)*, vol. 4, pp. 356–357.

Reeves, F., and R. O. Chalmers. 1949. "The Wolf Creek Crater." *Austral. Jour. Sci.* (April), pp. 154–156.

Reiff, W., ed. 1979. *Guidebook to the Steinheim Basin Impact Crater.* Geologisches Landesamt Baden-Wurttemberg, West Germany.

Reinvaldt, I. A. 1933. *Kaali Järv—The Meteorite Craters on the Island of Ösel (Estonia).* Publications of the Geological Inst. of the Univ. of Tartu, no. 30, Tallinn, Estonia.

———. 1939. *The Kaalijärv Meteor Craters (Estonia), Supplementary Research of 1937: Discovery of Meteoric Iron.* Publications of the Geological Inst. of the Univ. of Tartu, no. 55, Tallinn, Estonia.

Rhodes, R. C. 1975. "New Evidence for Impact Origin of the Bushveld Complex, South Africa." *Geology* (October), pp. 549–554.

Rhodes, R. C., and M. D. Du Plessis. 1975. "Notes on Some Stratigraphic Relations in the Rooiberg Felsite." *Trans. Geol. Soc. South Africa*, vol. 79, part 2, pp. 183–185.

Rinehart, J. S. 1957. "Distribution of Meteorite Debris Around the Arizona Meteorite Crater." *Astron. Jour.*, vol. 62, p. 96.

———. 1958. "Impact Effects and Tektites." *Geochim. et Cosmochim. Acta*, vol. 14, pp. 287–290.

Ringwood, A. E. 1960. "Cohenite as a Pressure Indicator in Iron Meteorites." *Geochim. et Cosmochim. Acta*, vol. 20, pp. 155–157.

Robertson, P. B. 1968. "La Malbaie Structure, Quebec—A Paleozoic Meteorite Impact Site. *Meteoritics*, vol. 4, pp. 89–112.

Rogers, A. F. 1930. "A Unique Occurrence of Lechatelierite or Silica Glass." *Am. Jour. Sci.*, series 5, vol. 19, pp. 195–202.

Rohleder, H. P. T. 1933. "The Steinheim Basin and the Pretoria Salt Pan—Volcanic or Meteoric Origin?" *Geol. Mag.*, vol. 70, pp. 489–498.

———. 1936. "Lake Bosumtwi, Ashanti." *Geograph. Jour.*, vol. 87, pp. 51–65.

Rouse, J. T., and C. H. Behre, Jr. 1966. "Memorial to Walter Herman Bucher (1888–1965)." *Bull. Geol. Soc. Am.*, vol. 77, part I, p. P99.

Russell, I. C. 1885. "Geologic History of Lake Lahontan, a Quaternary Lake of N.W. Nevada." *U.S.G.S. Monograph No. 11*, p. 73.

Sandberg, C. 1907. "Notes on the Structural Geology of South Africa." *Trans. Inst. Min. Eng.*, vol. 33, pp. 540–557.

Sander, B. 1930. *Gefügekunde der Gesteine.* Julius Springer, Vienna.

Schnetzler, C. C., W. H. Pinson, and P. M. Hurley. 1966. Rubidium-Strontium Age of the Bosumtwi Crater Area, Ghana, Compared with the Age of the Ivory Coast Tektites." *Science*, vol. 151, pp. 817–819.

Schuchert, C. 1930. "George Perkins Merrill, 1854–1929." *Smithsonian Inst. Ann. Rept.*, pp. 617–633.

Schwarz, E. H. L. 1909. "The Probability of Large Meteorites Having Fallen on the Earth." *Jour. Geol.*, vol. 17, pp. 124–135.

Sears, D. W. 1975. "Sketches in the History of Meteoritics I: The Birth of the Science." *Meteoritics*, vol. 10, pp. 215–225.

————. 1976. "Edward Charles Howard and an Early British Contribution to Meteoritics." *Brit. Astron. Assoc. Jour.*, vol. 86, pp. 133–139.

See, T. J. J. 1910. "The Origin of the So-Called Craters on the Moon by the Impact of Satellites, and the Relation of these Satellite Indentations to the Obliquities of the Planets." *Pubn. Astron. Soc. of the Pacific*, vol. 22, pp. 13–20.

Sellards, E. H., and G. Evans. 1940. "Statement of Progress of Investigation at Odessa Meteor Craters." Univ. of Texas Bureau of Econ. Geol. (Sept. 1), mimeographed.

Shand, S. J. 1916. "The Pseudotachylite of Parijs (O.F.S.) and Its Relation to 'Trap-Schotten Gneiss' and 'Flinty Crush-Rock'." *Quart. Jour. Geol. Soc. of London*, vol. 72, pp. 198–221.

————. 1928. "The Geology of Pilansberg (Pilaan's Berg) in the Western Transvaal: A Study of Alkaline Rocks and Ring-intrusions." *Trans. Geol. Soc. South Africa*, vol. 31, pp. 97–156.

————. 1930. "Limestone and the Origin of Felspathoidal Rocks: An Aftermath of the Geological Congress." *Geol. Mag.*, vol. 67, pp. 415–426.

Shoemaker, E. M. 1956. "Occurrence of Uranium in Diatremes on the Navajo and Hopi Reservations, Arizona, New Mexico and Utah." *U.S.G.S. Prof. Paper No. 300*, pp. 179–185.

————. 1957. "Primary Structures of Maar Rims and Their Bearing on the Origin of Kilbourn Hole and Zuni Salt Lake, New Mexico" (abs.). *Bull. Geol. Soc. Am.*, vol. 68, p. 1846.

————. 1960a. "Penetration Mechanics of High Velocity Meteorites, Illustrated by Meteor Crater, Arizona." Report of the 21st International Geological Congress, Norden, Part XVIII, *Structure of the Earth's Crust and Deformation of Rocks*, edited by A. Kvale and A. Metzger, pp. 418–434. Copenhagen.

————. 1960b. "Brecciation and Mixing of Rock by Strong Shock." *U.S.G.S. Prof. Paper No. 400B*, pp. B423–B425.

————. 1962a. "Interpretation of Lunar Craters." In *Physics and Astronomy of the Moon*, edited by Zdenek Kopal, pp. 283–359. Academic Press, New York.

————. 1962b. "Exploration of the Moon's Surface." *Am. Scientist*, vol. 50, pp. 99–130.

————. 1963. "Impact Mechanics at Meteor Crater, Arizona." In *The Moon, Meteorites, and Comets*, edited by B. M. Middlehurst and G. P. Kuiper, pp. 301–336. Univ. of Chicago Press.

————. 1976. "Why Study Impact Craters?" In *Impact and Explosion*

*Cratering,* edited by D. J. Roddy, R. O. Pepin, and R. B. Merrill, pp. 1–10. Pergamon Press, New York.

Shoemaker, E. M., and E. C. T. Chao. 1961. "New Evidence for the Impact Origin of the Ries Basin, Bavaria, Germany." *Jour. Geophys. Res.,* vol. 66, pp. 3371–3378.

Shoemaker, E. M., D. E. Gault, and R. V. Lugn. 1961. "Shatter Cones Formed by High Speed Impact in Dolomite." *U.S.G.S. Prof. Paper No. 424D,* no. 147, pp. D365–D368.

Shoemaker, E. M., and K. E. Herkenhoff. 1983. "Impact Origin of Upheaval Dome, Utah." *EOS,* vol. 64, part 2, p. 747.

Short, N. M. 1966. "Shock Processes in Geology." *Jour. Geol. Education,* vol. 14, pp. 149–166.

———. 1970. "Evidence and Implications of Shock Metamorphism in Lunar Samples." *Science,* vol. 176, pp. 673–675.

Short, N. M., and T. E. Bunch, 1968. "A Worldwide Inventory of Features Characteristic of Rocks Associated with Presumed Meteorite Impact Structures." In *Shock Metamorphism of Natural Materials,* edited by B. M. French and N. M. Short, pp. 255–266. Mono Book Corp., Baltimore.

Shrock, R. R., and C. A. Malott. 1930. "Notes on Some Northwestern Indiana Rock Exposures." *Proc. Indiana Acad. Sci. for 1929,* vol. 39, pp. 221–227.

———. 1933. "The Kentland Area of Disturbed Ordovician Rocks in Northwestern Indiana." *Jour. Geol.,* vol. 41, pp. 337–370.

Silliman, B., and J. L. Kingsley. 1809. "An Account of the Meteor, Which Burst over Weston in Connecticut, in December 1807, and of the Falling of Stones on that Occasion." *Proc. Am. Phil. Soc.,* no. 15, pp. 141–161.

Simpson, E. S. 1938. "Some New and Little-known Meteorites Found in Western Australia." *Mineralog. Mag.,* vol. 25, pp. 157–170.

Skinner, B. J., and J. J. Fahey. 1963. "Observations on the Inversion of Stishovite to Silica Glass." *Jour. Geophys. Res.,* vol. 68, pp. 5595–5604.

Skrine, C. P. 1931. "The Highlands of Persian Baluchistan." *Geograph. Jour.,* vol. 78, pp. 321–340.

Smith, C. S. 1962. "Note on the History of the Widmanstätten Structure." *Geochim. et Cosmochim. Acta,* vol. 26, pp. 971–972.

Smith, W. C. 1952. "L. J. Spencer's Work at the British Museum." *Mineralog. Mag.,* vol. 29, pp. 256–270.

Smyth, J. R., and C. J. Hatton. 1977. "Coesite-Sanidine Grospydite from the Roberts Victor Kimberlite." *Earth and Planetary Sci. Letters,* vol. 34, pp. 284–290.

Snyder, F. G., and P. E. Gerdemann. 1965. "Explosive Igneous Activity Along an Illinois-Missouri-Kansas Axis." *Am. Jour. Sci.,* vol. 263, pp. 465–493.

Speers, E. C. 1957. "The Age Relation and Origin of Common Sudbury Breccia." *Jour. Geol.,* vol. 65, pp. 497–514.

Spencer, L. J. 1930. "Meteoric Irons from Southwest Africa." *Nat. Hist. Mag.* (British Museum), vol. 2, pp. 240–246.

————. 1932a. "Hoba (South-West Africa), the Largest Known Meteorite." *Mineralog. Mag.*, vol. 23, no. 136, pp. 1–18.

————. 1932b. "Meteorite Craters." *Nature*, vol. 129, pp. 781–784.

————. 1933a. "Meteoric Iron and Silica Glass from the Meteorite Craters of Henbury (Central Australia) and Wabar (Arabia)." *Mineralog. Mag.*, vol. 23, pp. 387–404.

————. 1933b. "Meteorite Craters as Topographic Features on the Earth's Surface." *Geograph. Jour.*, vol. 81, pp. 227–248.

————. 1933c. "Meteorites and Fulgurites." Appendix A in *The Empty Quarter*, by H. St. J. Philby, pp. 365–370. Constable & Co., London.

————. 1933d. "Origin of Tektites." *Nature* (January 26), pp. 117–118.

————. 1933e. "Origin of Tektites (Reply to Chas. Fenner)." *Nature*, vol. 132 (October 7), p. 571.

————. 1937. "Meteorites and Craters on the Moon." *Nature* (April 17), pp. 655–657.

————. 1941. "The Gibeon Shower of Meteoritic Irons in South-West Africa." *Mineralog. Mag.*, vol. 26, pp. 19–35.

Spurr, J. E. 1944. *Geology Applied to Selenology*. Science Press Printing Co., Lancaster, Penn.

Stevenson, J. S. 1960. "Origin of Quartzite at the Base of the Whitewater Series, Sudbury Basin, Ontario." *21st International Geological Congress Proc.*, Norden, sec. 13, part 26, pp. 32–41. Copenhagen.

Stishov, S. M., and S. V. Popova. 1961. "A New Dense Modification of Silica." *Geikhimiya*, vol. 10, pp. 923–926.

Stöffler, D. 1966. "Zones of Impact Metamorphism in the Crystalline Rocks of the Nördlinger Ries Crater." *Contr. Mineral. and Petrol.*, vol. 12, pp. 15–24.

Storzer, D., and G. A. Wagner, 1979. "Fission Track Dating of El-gygytgyn, Popigai and Zhamanshin Impact Craters: No Source for Australasian or North American Tektites." *Meteoritics*, vol. 14, p. 541.

Suess, F. E. 1936. "Der Meteor-Krater von Köfels bei Umhausen im Ötztale, Tirol." *Neues Jahrbuch fhur Mineralogie, Geologie, un Palhaontologie*, vol. A, no. 72, pp. 98–155.

Svensson, N. B., and F. E. Wickman. 1965. "Coesite from Lake Mien, Southern Sweden." *Nature*, vol. 205, pp. 1202–1203.

Taylor, S. R., and M. Solomon. 1964. "The Geochemistry of Darwin Glass." *Geochim. et Cosmochim. Acta*, vol. 28, pp. 471–494.

Thomas, R. N., and F. L. Whipple. 1951. "The Physical Theory of Meteors II. Astroballistic Heat Transfer." *Astrophys. Jour.*, vol. 114, pp. 448–465.

Thompson, M. 1892. "Geology of Carrol County." Indiana Dept. of Geol. and Nat. Resources, *17th Ann. Rept.*, pp. 177–186.

Thomson, E. 1914. Letter No. 4 to D. M. Barringer. In *Selections from the Scientific Correspondence of Elihu Thomson*, edited by H. J. Abrahams and M. B. Savin. MIT Press, Cambridge, Mass., 1971.

Thomson, J. E. 1957. "Recent Geological Studies in Sudbury Camp." *Can. Min. Jour.*, vol. 78, pp. 109–112.

Thornbury, W. D. 1965. *Regional Geomorphology of the United States.* John Wiley & Sons, New York.

Thorndike, L. 1958. *History of Magic and Experimental Science.* Vol. 8. Columbia Univ. Press.

Tilghman, B. C. 1905. "Coon Butte, Arizona." *Proc. Acad. Nat. Sci. Phil.,* vol. 57, pp. 887−914.

Tilley, C. E. 1961. "Leonard James Spencer." *Biographical Memoirs of Fellows of the Royal Society,* vol. 7, pp. 243−248.

Tilton, G. R. 1958. "Isotopic Composition of Lead from Tektites." *Geochim. et Cosmochim. Acta,* vol. 14, pp. 323−330.

Tschermak, M. G. 1892. "Die Meteoriten von Shergotty und Goalpur." *Sitzungsber 1 Akad. Wiss. Wien. Math- Naturwiss.* K1, 65, part 1, pp. 122−146.

Turkevich, A. L., E. J. Franzgrote, and J. H. Patterson. 1969. "Chemical Composition of the Lunar Surface in Mare Tranquillitatis." *Science.* vol. 165, pp. 277−279.

Turner, F. J., and C. S. Ch'ih. 1951. "Deformation of Yule Marble: Part III — Observed Fabric Changes Due to Deformation at 10,000 Atmospheres Confining Pressure, Room Temperature, Dry." *Bull. Geol. Soc. Am.,* vol. 62, pp. 887−906.

Turner, F. J., D. T. Griggs, R. H. Clark, and R. H. Dixon. 1956. "Deformation of Yule Marble, Part VII—Development of Oriented Fabrics at 300°C−500°C." *Bull. Geol. Soc. Am.,* vol. 67, pp. 1259−1294.

Turner, F. J., D. T. Griggs, H. Heard, and L. W. Weiss. 1954. "Plastic Deformation of Dolomite Rock at 380°C." *Am. Jour. Sci.,* vol. 252, pp. 477−488.

Uhlig, H. H. 1954. Contribution of Metallurgy to the Origin of Meteorites, Part I— Structure of Metallic Meteorites, Their Composition and the Effect of Pressure." *Geochim. et Cosmochim. Acta,* vol. 6, pp. 282−301.

Urey, H. C. 1956a. "Diamonds, Meteorites, and the Origin of the Solar System." *Astrophys. Jour.,* vol. 124, pp. 623−637.

———. 1956b. "The Origin of the Moon's Surface Features. I and II." *Sky and Telescope* (January), pp. 108−111, (February), p. 161.

———. 1957. "Origin of Tektites." *Nature,* vol. 179, pp. 556−557.

———. 1962. "Origin of Tektites." *Science,* vol. 137, pp. 746−748.

Urey, H. C., A. Mele, and T. Mayeda. 1957. "Diamonds in Stone Meteorites." *Geochim. et Cosmochim. Acta,* vol. 13, pp. 1−4.

Vauquelin, L. N. 1802. "Sur les pierres dite tombées du ciel." *Ann. de Chim. et de Phys.,* vol. 46, pp. 225−245.

Vinogradov, A. P., and G. P. Vdovykin. 1963. "Diamonds in Stony Meteorites." *Geokhimiya,* no. 8, pp. 715−720.

von Bieberstein, M. 1802. *Untersuchungen Über den Ursprung und die Ausbildung der gegenwartigen Anordnung des Weltgebaeudes.* Darmstadt.

Wackerle, J. 1962. "Shock-Wave Compression of Quartz." *Jour. Applied Physics,* vol. 33, pp. 922−937.

Walker, R. T. 1928. "Mineralized Explosion Pipes." *Eng. and Min. Jour.,* vol. 126, pp. 895–898, 939–941, and 976–984.

Walker, T. L. 1897. "Geological and Petrographical Studies of the Sudbury Nickel District (Canada)." *Quart. Jour. Geol. Soc. London,* vol. 53, pp. 40–66.

Wallis, J. 1677. "Extract of Three Letters of Dr. Wallis Concerning an Unusual Meteor Seen at the Same Time in Many Different Places in England." *Phil. Trans. Roy. Soc.* no. 135.

Walter, L. S. 1965. "Coesite Discovered in Tektites." *Science,* vol. 147, pp. 1029–1032.

Ward, L. C. 1906. "Roads and Road Materials of the Northern Third of Indiana." Indiana Dept. Geol. and Nat. Resources, *30th Ann. Rept.,* pp. 214–219.

Wegener, A. 1921. "The Origin of Lunar Craters" (translated by A. M. Celâl Sengör. *The Moon,* vol. 14, pp. 211–236, 1975.

Wentorf, R. N., Jr., and H. P. Bovenkerk. 1961. "On the Origin of Natural Diamonds." *Astrophys. Jour.,* vol. 134, pp. 995–1005.

Werner, E. 1904. "Das Ries in der Schwäbisch-frankischen Alb." *Blatter der Schwab., Albvereins,* vol. 16, pp. 153–167.

Whipple, F. J. W. 1930. "The Great Siberian Meteor and the Waves, Seismic and Aerial, Which It Produced." *Royal Meteorological Soc. Quart. Jour.,* vol. 56, pp. 287–304.

Whipple, F. L. 1972. "Introduction." *Find a Falling Star,* by H. H. Nininger. Paul S. Eriksson, Inc., New York.

Willemse, J. 1964. "A Brief Outline of the Geology of the Bushveld Igneous Complex." In *The Geology of Some Ore Deposits in Southern Africa,* edited by S. H. Haughton, vol. 2, pp. 92–128. Geological Society of South Africa, Johannesburg.

———. 1969. "The Geology of the Bushveld Igneous Complex, the Largest Repository of Magmatic Ore Deposits in the World." *Economic Geology Monograph No. 4.*

Williams, G. H. 1891. "The Silicified Glass-Breccia of Vermillion Ridge, Sudbury District." *Bull. Geol. Soc. Am.,* vol. 2, pp. 138–140.

Williams, H. 1956. "Glowing Avalanche Deposits of the Sudbury Basin." *65th Ann. Rept. of the Ontario Dept. Mines,* vol. 65, part 3, pp. 57–89.

Willmore, P. L., and A. E. Scheidegger. 1956. "Seismic Observations in the Gulf of St. Lawrence." *Trans. Roy. Soc. Can.,* vol. 50, series 3, pp. 21–38.

Wilshire, H. G. 1971. "Pseudotachylite from the Vredefort Ring, South Africa." *Jour. Geol.,* vol. 79, pp. 195–206.

Wilson, C.W. 1953. "Wilcox Deposits in Explosion Craters, Stewart County, Tennessee, and Their Relations to the Origin and Age of Wells Creek Basin Structure." *Bull. Geol. Soc. Am.,* vol. 64, pp. 753–768.

Wilson, C. W., and K. E. Born. 1936. "Flynn Creek Disturbance at Jackson, Tennessee." *Jour. Geol.,* vol. 44, pp. 815–835.

Yates, A. B. 1938. "The Sudbury Intrusive." *Trans. Roy. Soc. Can.*, vol. 32 (4), pp. 151–172.

———. 1948. "Properties of International Nickel Company of Canada." In *Structure and Geology of Canadian Ore Deposits: A Symposium.* Can. Inst. of Min. and Metallurg., pp. 596–617.

# Illustration Credits

| | |
|---|---|
| Frontispiece, figures 2.3, 3.1, and 3.2 | Photographs by John S. Shelton. All rights reserved. |
| Figure 1.1 | From F. A. Paneth (1960, fig. 1); reprinted by permission. |
| Figure 1.2 | From F. A. Paneth (1960, fig. 4); reprinted by permission. |
| Figure 2.1 | From V. C. Kelley (1956, p. 171). |
| Figure 2.2 | From G. K. Gilbert (1877, figs. 7–9). |
| Figure 2.4 | From *Geology Illustrated*, by John Shelton; copyright © 1966 by W. H. Freeman and Company. All rights reserved. |
| Figure 2.5 | From J. B. Dawson, 1967, "A Review of the Geology of Kimberlite," in *Ultramafic and Related Rocks*, edited by P. J. Wyllie, p. 247. Copyright © by John Wiley & Sons, Inc. Reprinted by permission of J. B. Dawson and John Wiley & Sons, Inc. |
| Figures 4.1, 4.2, and 4.3 | From E. L. Krinov (1966, figs. 157, 159, 129); reprinted by permission. |
| Figure 4.4 | Photograph by R. F. Fudali, Smithsonian Institution. |
| Figure 6.1 | Based on "Decade of North American Geology, 1983, Geologic Time Scale." Geological Society of America. |
| Figure 6.2 | Courtesy of the National Film Board of Canada, and R. S. Dietz. |
| Figure 6.3 | Adapted from Cumings and Shrock (1927, p. 81); used by permission. |
| Figure 6.4 | Adapted from J. D. Boon and C. C. Albritton, Jr. (1937, fig. 2); used by permission. |
| Figures 7.1, 7.2, and 7.3 | Courtesy of the American Meteorite Laboratory |
| Figure 7.4 | Photograph by Joan Neary. |

| Figures 8.1, 10.5, 10.7, and 14.8 | Photographs by R. S. Dietz. |
|---|---|
| Figure 8.2 | Courtesy of NASA (photograph S84-29410). |
| Figure 8.3 | Photograph no. T193R-87. Copyright © 1948 Her Majesty the Queen in Right of Canada, reproduced from the collection of the National Air Photo Library with permission of Energy, Mines and Resources Canada. |
| Figure 8.4 | From V. B. Meen (1950, fig. 2); drawn by G. F. McCauley. Courtesy of the Royal Ontario Museum. |
| Figure 8.5 | From E. L. Krinov (1966, fig. 181); painting by P. I. Medvedjev. Reprinted by permission. |
| Figure 10.1 | From E. C. T. Chao, E.M. Shoemaker, and B. M. Madsen, 1960, "First Natural Occurrence of Coesite," *Science,* vol. 132, p. 221, fig. 3. Copyright © 1960 by the American Association for the Advancement of Science; reprinted by permission. |
| Figure 10.2 | Photograph by Kathleen Mark. |
| Figure 10.3 | From E. M. Shoemaker and E. C. T. Chao, 1961, "New Evidence for the Impact Origin of the Ries Basin, Bavaria, Germany," *Jour. Geophys. Res.,* vol. 66, p. 3372, copyright © 1961 by the American Geophysical Union. Reprinted by permission. |
| Figure 10.4 | Photograph by W. M. Woodhouse. Courtesy of R. S. Dietz. |
| Figure 10.6 | Photograph courtesy of A. Brugger, Stuttgart, December 1968. Ltd. of Regierungspräsidium Nordwürttemberg. No. 2/27814. |
| Figures 11.1, 11.2, 11.3, 11.4, and 12.4 | Courtesy of the Earth Physics Branch, Department of Energy, Mines and Resources, Ottawa, Canada |
| Figure 11.5 | From E. M. Shoemaker (1960a, fig. 4); reprinted by permission of the author. |
| Figure 11.6 | After E. H. Sellards and B. Evans, 1941; used by permission. |
| Figure 11.7 | From E. M. Shoemaker (1960a, fig.8); reprinted by permission of the author. |
| Figure 12.1 | From F. G.Snyder and P. E. Gerdemann, 1965, "Explosive Igneous Activity Along an Illinois-Missouri-Kansas Axis." *Am. Jour. Sci,* vol. 263, p. 467. Reprinted by permission of |

|  | P. E. Gerdemann and the *American Journal of Science.* |
|---|---|
| Figure 12.2 | Courtesy of NASA (photograph S09-48-3139). |
| Figure 12.3 | From P. L. Willmore and A. E. Scheidegger (1956, p. 21); reprinted by permission. |
| Figure 12.5 | From M. R. Dence, 1965, "The Extraterrestrial Origin of Canadian Meteorite Craters." *Annals N. Y. Acad. Sci.,* vol. 123, p. 943. Reprinted by permission of M. R. Dence and the New York Academy of Sciences. |
| Figure 12.6 | From B. M. French (1970, p. 471; adapted from Dence, 1968); reprinted by permission of the author. |
| Figure 13.1 | After D. Griggs, F. J. Turner, I. Borg, and J. Sosoka, 1953, "Deformation of Yule Marble: Part V—Effects at 300°C." *Geol. Soc. Am. Bull.,* vol. 64, pp. 1327–1342, pl. 1. Courtesy of the Bancroft Library, University of California, Berkeley. Photography by J. Hampel, 1984. Used by permission of F. J. Turner. |
| Figure 13.2 | From R. Berman and F. Simon (1955, p. 336); used by permission. |
| Figure 14.2 | From J. Willemse (1969, frontispiece); reprinted by permission. |
| Figure 14.3 | Courtesy of NASA (photograph S08-33-1063). |
| Figure 14.4 | Courtesy of NASA (photograph S08-37-1937). |
| Figure 14.5 | Courtesy of NASA (photograph S08-31-0771). |
| Figure 14.6 | After A. L. Hall and G. A. F. Molengraaff (1925, p. 1); used by permission. |
| Figure 14.7 | From S. J. Shand (1916, pl. XVII); reprinted by permission of Cambridge University Press. |
| Figure 14.9 | From R. C. Rhodes, 1975, *Geology* (October), p. 549 (courtesy of Geological Society of America). |
| Figure 15.2 | From A. P. Coleman (1907, pp. 761, 763). |
| Figure 15.3 | From R. S. Dietz and L. W. Butler, 1964, "Shatter-cone Orientation at Sudbury, Canada." *Nature,* vol. 204, no. 4955, p. 281. Reprinted by permission. Copyright © 1964, Macmillan Journals Limited. |
| Figure 16.1 | From C. Darwin (1896, p. 44). |
| Figure 16.2 | From H. H. Nininger (1943, p. 12); reprinted by permission of the author. |

# Index

Abaskansk, 8
Abercrombie, J. T., 49
Accretion, 107, 174, 234
Adams, F. D., 167–68
Adrar desert, 92, 94
Ahnigito (meteorite), 75, 94
Albaretto, 5
Albritton, C. C. Jr. See Boon, J. D., and
    C. C. Albritton, Jr.
Alderman, A. R., 45, 54, 81
American Association for the
    Advancement of Science, 34
American Chemical Society, 174
American Geographical Society, 31
American Geophysical Union, 153
American Museum of Natural History, 75
Amstutz, G. C., 156
Anders, E., 116, 173–77
Aouelloul, crater of, 91–92, 94, 101, 177
Arizona crater. See Meteor Crater
Ashanti crater. See Bosumtwi (Lake)
    crater
Assen ring, 203
Aureoles. See Bushveld Igneous
    Complex, Vredefort
Avicenna, 8
Avon region, 154

Bachelay, Abbe, 5, 11
Bacubirito (iron meteorite), 75
Baddelyite, 177
Bailey, E. B., 197
Baker, Marcus, 27
Baldwin, R. B., 107–9, 129, 132, 199
Banks, Sir Joseph, 10
Barlow, A. E., 208
Barnes, V. E., 120, 230, 232
Barringer, D. M., 31–38, 40, 78, 84, 88,
    106, 108, 113, 122, 165
    examination of Meteor Crater, 31–33
Barringer, D. M., Jr., 40–41
Barringer, J. P., 228
Barringer Crater, 113, 148–49, 151.
    See also Meteor Crater
Barringer Crater Company, 116. See
    also Standard Iron Company
Barringer Meteorite Crater, 25, 113. See
    also Meteor Crater

Beals, C. S., 98, 130, 133–34, 136,
    138–39, 147, 151, 153, 156, 158,
    160, 164
    forms of ancient craters, 138–39
    search for fossil meteorite craters,
    133–34, 136, 138, 153
Beard, D. P., 106
Bedford, R., 55
Beer, W., 104
Bell, J. F., 168
Bell, Robert, 210
Benares meteoritic stones, 10–11
Berman, R., 171
Bibbins, A. B., 41, 81
Bickerton, A. W., 51
Bilateral symmetry, 36, 72–73, 90, 200
Biot, J. B., 13
Birch, Francis, 166
Bishopp, D. W., 198
Bjork, R. L., 147
Black Pearls. See Silica glass
Blackwelder, Eliot, 114
Blagden, Charles, 5
Bohemia meteoritic stones, 9, 11
Bolides, 2, 6, 15, 98, 224. See also
    Meteors
Bonney, T. G., 210, 233
Boon, J. D., and C. C. Albritton Jr.,
    71–73, 107, 110, 111–12, 123,
    127, 137, 179, 197–98, 200
    interpretation of cryptovolcanic
    structures, 71–73; of Vredefort,
    197–98
Born, K. E., 70, 73
Bosumtwi (Lake) crater, 47, 48, 100–1,
    120–21, 124, 142, 232
Bournon, Count de, 11
Bovenkerk, H. P., 173, 175
Boxhole crater, 90, 100–1
Boyd, F. R., 117
Boyle, Robert, 165
Branca (Branco) W., 62, 66, 149
Branner, J. C., 34, 36
Bray, J. Guy, 222, 224, 225
Breccia, 64, 66, 71, 98, 117, 119, 131,
    132, 138, 146, 161, 162, 163, 164,
    204, 208, 210, 213, 216–17, 220,
    223–25, 234